ACCUMULATION OF NITRATE

Committee on Nitrate Accumulation
Agricultural Board
Division of Biology and Agriculture
National Research Council

NATIONAL ACADEMY OF SCIENCES
Washington, D.C.
1972

NOTICE: The study reported herein was undertaken under the aegis of the National Research Council with the express approval of its Governing Board. Such approval indicated that the Board considered the problem to be of national significance, that its elucidation required scientific or technical competence, and that the resources of NRC were particularly suitable to the conduct of the project. The institutional responsibilities of the NRC were then discharged in the following manner:

The members of the study committee were selected for their individual scholarly competence and judgment with due consideration for the balance and breadth of disciplines. Responsibility for all aspects of this report rests with the study committee, to whom sincere appreciation is expressed.

Although the reports of our study committees are not submitted for approval to the Academy membership or to the Council, each report is reviewed by a second group of scientists according to procedures established and monitored by the Academy's Report Review Committee. Such reviews are intended to determine, *inter alia*, whether the major questions and relevant points of view have been addressed and whether the reported findings, conclusions, and recommendations arose from the available data and information. Distribution of the report is approved, by the President, only after satisfactory completion of this review process.

This study was supported by the Environmental Protection Agency and the U.S. Department of Agriculture.

Available from

Printing and Publishing Office
National Academy of Sciences
2101 Constitution Avenue, N.W.
Washington, D.C. 20418

ISBN 0-309-02038-7
Library of Congress Catalog Card Number 72-84111

Printed in the United States of America

Preface

The Committee on Nitrate Accumulation was established to examine various problems associated with the accumulation of nitrate nitrogen and related nitrogenous compounds in the environment and to recommend courses of action that may mitigate these problems.

Where possible, the Committee made specific recommendations. In some areas, however, lack of information prevented the Committee from suggesting definite courses of action. For example, information is lacking on the significance of various means of controlling nitrogen in waterways, on methods for reducing or increasing the amount of nitrogen lost from the soil, on the importance of nitrosamines in nature, and on whether "subclinical" hazards arise from the consumption of food and water containing small amounts of nitrate.

The Committee recommended research on various aspects of nitrogen as a fertilizer, food constituent, food additive, food preservative, and waste component.

Committee on Nitrate Accumulation

MARTIN ALEXANDER, *Chairman*, Cornell University

THOMAS J. ARMY, Great Western Sugar Company

FREDERICK J. deSERRES, Oak Ridge National Laboratory

CHARLES R. FRINK, Connecticut Agricultural Experiment Station

VICTOR J. KILMER, Tennessee Valley Authority

THURSTON E. LARSON, Illinois State Water Survey

NORTON NELSON, New York University Medical Center

W. H. PFANDER, University of Missouri

GERARD A. ROHLICH, University of Wisconsin

PERRY R. STOUT, University of California

SYLVAN H. WITTWER, Michigan State University

Contents

Introduction

Nitrogen occupies a unique place in man's life. Although the element is abundant and essential for all living things, supplies of the forms available to plants are inadequate in many parts of the world. Where the inadequacy exists, crop production is limited, and harvests are often insufficient to meet the needs of the people.

Early in history, man learned to compensate for the limited supply of nitrogen in many soils by using animal and human manure and by cultivating legumes. In recent times, he has been able to meet the large and constantly growing requirement by using chemical fertilizers.

Some forms of the element are hazardous to man or to the animals he raises for food. In certain concentrations, nitrate (and the nitrite derived from it) may be deleterious to the health of infants and livestock because of reactions these anions undergo in the body. Yet it has been found advantageous to add nitrate and nitrite to some food products. Despite the importance of nitrate as a nitrogen source for terrestrial plants and aquatic vegetation essential for fish production, increased availability of nitrogen also stimulates growth of algae in surface waters, which may adversely affect water quality and use. It has also been suggested that nitrite may enter into organic combination with amines to yield compounds, such as nitrosamines, that can be toxic to man and animals at very low concentrations.

Difficult problems arise from the unquestioned need for large amounts of fertilizer nitrogen, the possible hazards of environmental deterioration associated with the appearance in water of nitrate derived from fertilizer sources, and the hazards that may result from adding nitrate and nitrite to food products.

Nitrogen Compartments in the Biosphere

THE NITROGEN CYCLE

The usual description of the reactions of nitrogen in the biosphere involves the familiar concept of the nitrogen cycle. Nitrogen is visualized as moving in cyclic fashion from the atmosphere to the soil, where it is taken up by plants. The plants are eaten by animals and humans, and the nitrogen from all these tissues is ultimately returned to the atmosphere by microbial degradation of nitrogenous organic wastes and reduction of nitrate to nitrogen gas. However, nitrogen may accumulate temporarily in various parts of the biosphere because of differences in rates of the many reactions involved. The nitrogen cycle can be visualized more accurately by thinking of pools of nitrogen in a number of compartments within the biosphere and of exchanges of nitrogen among compartments. Such exchanges are controlled by rates of biological and chemical reactions and by hydrological transport (Figure 1).

Nitrogen as the inert gas N_2 constitutes nearly 80 percent of the volume of the earth's atmosphere. The original source of this nitrogen, before life appeared, is considered to have been a primeval atmosphere containing ammonia and nitrogen from the earth itself.

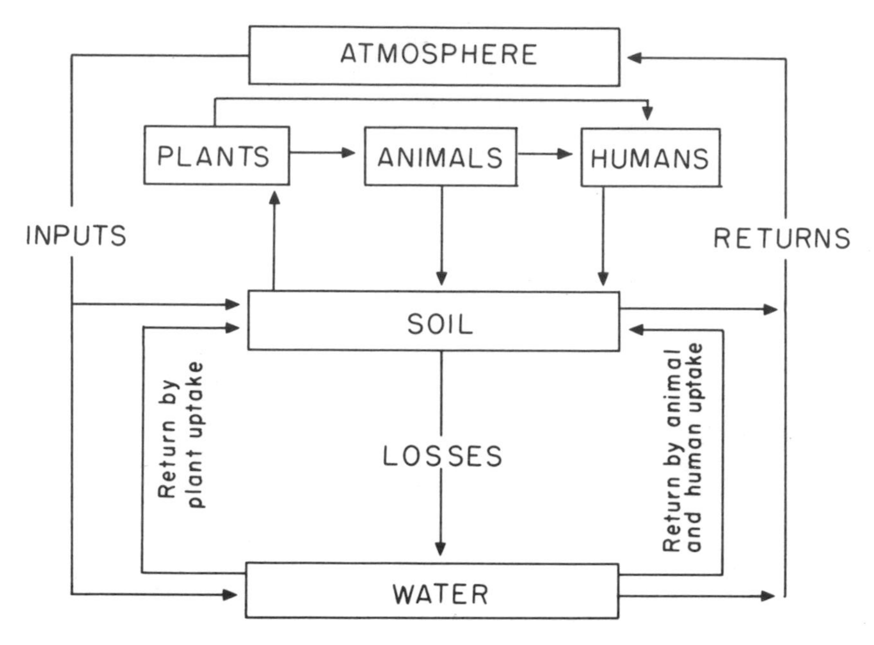

FIGURE 1 Nitrogen compartments in the environment.

As life evolved, the ammonia was converted by biological and chemical reactions to organic nitrogenous compounds. Other biological and chemical reactions restored free nitrogen to the atmosphere. The net storage of fixed nitrogen continued to a point where perhaps 10 billion metric tons of nitrogen had accumulated in soils of the United States before they were cultivated. This inherited biological accumulation is impressive when compared with the present chemical nitrogen fixation capacity in the United States, which is about 10 million metric tons annually.

Degradation of organic nitrogen compounds from dead and decaying organisms releases this fixed nitrogen to compartments other than the atmosphere. The reactions releasing nitrogen to the environment are complex, and a variety of processes are involved. Some microorganisms return nitrogen directly to the atmosphere as inert nitrogen gas. Other kinds of microbial reactions and chemical combustion form oxidized nitrogen compounds that react with atmospheric moisture and return to the earth in rainfall, and this nitrogen is recycled by living organisms or is transported to the water compartment. Fixed nitrogen compounds entering the water compartment may be stored in sediments or returned to the atmosphere by biochemical processes analogous to those of the soil compartment.

Although it is difficult to estimate the amounts of nitrogen returned to the atmospheric storage compartment via these various pathways, it is reasonable to assume that throughout post-Cambrian geological history, nitrogen returns were about equivalent to inputs and that the nitrogen cycle approached an equilibrium governed by regional climates and localized moisture conditions. When climates changed, the sizes of the nitrogen compartments changed. Coal beds are the residues of former nitrogen compartments, and substantial amounts of nitrogen that were fixed in the carboniferous period are retained in coals.

Man has created a series of perturbations in the various processes involved in the nitrogen cycle. These man-made disturbances may cause accumulations and depletions of nitrogen in various compartments if the rate of nitrogen input differs from the rate of its return. For example, small nitrogen inputs into the open water compartment may either significantly enhance the growth of aquatic life or cause an increase in the concentration of nitrate in drinking water. Similarly, high concentrations of nitrate may appear in plants consumed by animals or man. These disturbances usually result from man's attempt to manage parts of the nitrogen cycle to his advantage; occasionally, however, they have undesirable consequences.

FORMS OF NITROGEN

Two gases and four forms of nongaseous or combined nitrogen are important in the nitrogen cycle. The gases are molecular nitrogen and nitrous oxide. The forms of nongaseous or combined nitrogen are the amino groups, ammonium, nitrite, and nitrate. The amino groups are constituents of plant and animal protein and are also found in soil organic matter. The positive ammonium ion is either released from proteinaceous organic matter or urea, or it is synthesized by industrial processes involving fixation of atmospheric nitrogen. The negative nitrite ion is formed from the nitrate or ammonium ion by certain microorganisms in soil, water, sewage, and the alimentary tract. The negative nitrate ion is formed by the complete oxidation of ammonium by microorganisms in soil or water. When ammonium ions are formed in or are added to soils, they are usually converted rapidly by these microorganisms to nitrate wherever temperatures, soil acidity, and aeration are conducive to plant growth.

Growing plants assimilate either nitrate or ammonium and convert these ions to protein. When soils containing nitrate become waterlogged, ubiquitous soil microorganisms consume the dissolved oxygen, thereby making the waterlogged soil anaerobic. Denitrification then takes place, and nitrate and nitrite are converted mainly to nitrogen gas or nitrous oxide gas, which escapes to the atmosphere.

The ammonium ion is relatively immobile in soils because the negative charge of the soil particles retains the positively charged ammonium ion. In contrast, nitrite and nitrate ions are negatively charged and are thus repelled by the negatively charged soil particles. Because of this repulsion phenomenon, nitrate moves freely out of the soil system in both leaching and runoff water.

NITROGEN BUDGETS

To determine whether man, in his effort to obtain food, is creating undesirable accumulations of nitrogen within the various atmospheric, soil, and water compartments that make up the total environment, it is necessary to examine sources of nitrogen in the environment and to determine whether there are discernible changes in concentrations of nitrogen in atmospheric, soil, or water compartments. Nitrogen sources may be evaluated in several ways. The most common is a nitrogen budget, which shows average annual inputs and returns. If averages are taken over the entire country, focal points in the system where nitrogen is accumulating may be obscured. Moreover, direct measurements of various sources are imperfect and can lead to erroneous conclusions. Nevertheless, preparation of budgets does serve useful purposes, including the identification of gaps in knowledge.

INPUTS

Direct inputs to the soil compartment are nitrogen fixed both by symbiotic associations of organisms and by free-living microorganisms, nitrogen fixed as fertilizer, and nitrogen compounds entering in rainfall or as fallout of particulate material. Estimates of these sources are shown in Table 1. With the exception of the input of about 7.5 million metric tons in chemical fertilizer (Farrell, 1971), these estimates are only approximate.

TABLE 1 Estimates of Nitrogen Inputs and Returns to the Total Land Area of the United States, 1970 (millions of metric tons of nitrogen)[a]

Annual Inputs to Soil Compartment	
Nonsymbiotic N_2 fixation	1.2
Symbiotic N_2 fixation	3.6
Rainfall	5.6
Chemical fixation	7.5
Mineralization of soil organic nitrogen	3.1
Total input	21.0
Utilization in Plant–Animal–Human Food and Fiber Chains	
Production of fiber	0.2
Production of sugar	0.6
Production of plant protein	0.9
Production of animal protein	15.1
Total	16.8
Total input	21.0
Not utilized in food chains	4.2
Fate of Nitrogen in Food and Fiber Chains	
Excreted by humans	1.2
Excreted by animals	4.2
Other nitrogenous wastes[b]	15.6
Total	21.0
Annual Returns to Atmospheric Compartment	
As ammonia or oxides to atmosphere moisture	5.6
By denitrification after loss to waterways	5.0
By denitrification from soil	8.9
Total	19.5
Total input	21.0
Net Retention in Soil and Water per Annum	1.5

[a] The values are only estimates.
[b] Total input minus that excreted by humans and animals.

A considerable amount of N_2 is fixed symbiotically by both wild and cultivated legumes. The amounts fixed by cultivated crops range from 50 to 500 kg nitrogen/ha; 50–100 kg nitrogen/ha annually has been reported as an average value (Allison, 1957). The quantity of nitrogen fixed by symbiotic reactions for the entire United States was estimated to be 1.8 million metric tons by Stanford *et al.* (1970), who assumed that for soybeans, clover, alfalfa, and grass–legume mixtures, one half of the nitrogen in crops harvested in 1969 was derived from symbiotic fixation of N_2. Because many of these crops receive little or no nitrogen fertilizer, the value of one half is probably low, and the estimate of symbiotic N_2 fixation in Table 1 has been doubled accordingly.

Inputs by nonsymbiotic fixation were estimated to be about 7 kg nitrogen/ha in the harvested crop area of the United States (Lipman and Conybeare, 1936) or about 0.9 million metric tons. Much higher values for specific crops have been observed. For example, Whitt (1941) calculated gains of 110 kg nitrogen/ha annually under bluegrass, and Parker (1957) reported gains of 70 to 80 kg nitrogen/ha under grass. Despite 35 years of research, the estimate of Lipman and Conybeare still seems appropriate, although it is probably somewhat low. This source is estimated to be about 1.2 million metric tons.

Estimates of total nitrogen input in rainfall in the United States vary from 1.4 to 7.1 million metric tons of nitrogen (Allison, 1965; Stanford *et al.*, 1970). Inputs of this magnitude indicate an average annual nitrogen contribution of 1.5–2.5 mg/liter to the upper 30 cm of a hectare of water. The Council on Environmental Quality (Train *et al.*, 1970) has estimated that about 7.4 million metric tons of nitrogen oxides were produced by transportation and 11.3 million tons from other combustion sources in the United States in 1968. If these oxides are assumed to contain 30 percent of nitrogen (as in NO_2), the potential return in rainfall from this source would be 5.6 million metric tons (Table 1). This figure neglects the biological production of oxides and any return of nitrogen in the form of ammonia. The nitrogen return as ammonia in rainfall generally exceeds that of nitrate by a factor of about two (Allison, 1965). Thus, our knowledge of these various inputs remains modest.

In addition to these direct inputs, several stored forms of nitrogen in the soil compartment are utilized by growing crops. These include so-called geologic nitrate (such as the Chilean nitrate deposits), nitrogen stored in soil organic matter, and nitrogen in the water compartment, which is tapped by deep-rooted crops or by irrigation. By far the largest source is the nitrogen in soil organic matter, which is released as soils are cultivated. The nitrogen content of virgin surface soils in the United States varied from 0.01 to 1 percent when first cultivated, 90 percent of which was contained in soil organic matter (Allison, 1957). Initially, cultivation results in a rather rapid decrease in soil nitrogen, but the rate of decrease declines as cultivation continues (Figure 2). Because the rate changes as cultivation is repeated, estimates of nitrogen from this source vary widely. Stauffer (1942) has shown that leaching losses of nitrogen from unfertilized fallow soils in Illinois may approach 85 kg nitrogen/ha per annum. For the total cropland area in the United States, the mineralization of soil organic nitrogen may release about 40 kg nitrogen/ha per annum, a

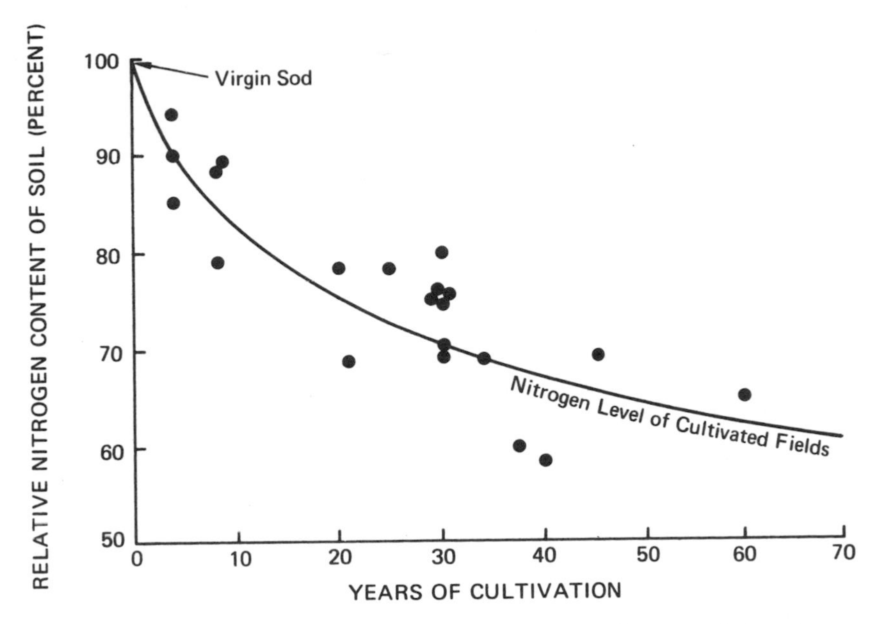

FIGURE 2 Decline of nitrogen content in soils with length of cultivation periods under average farming practices in the Midwest. From Jenny (1933).

total of 3.1 million metric tons. Thus, the total nitrogen made available for crop growth in the United States annually in the soil compartment is estimated to be 21 million metric tons (Table 1).

UTILIZATION IN PLANT–ANIMAL–HUMAN FOOD CHAINS

How are these sources of nitrogen in the soil compartment utilized in the plant–animal–human food chain? Humans in the United States consume about 1.2 million metric tons of nitrogen annually. About 70 percent is derived from animal protein and the remainder from plant protein. Since the conversion of soil nitrogen to plant protein is inefficient and the introduction of animal protein into the food chain adds even greater inefficiencies, the total amount of nitrogen required to produce this protein is much more than 1.2 million tons.

Estimates of the amounts of nitrogen necessary to produce plant and animal protein, sugar, and fiber are developed in a later chapter ("Fertilizer and Soil Nitrogen," p. 30) and are given in Table 1. In these calculations, the following levels of efficiency are assumed:

utilization of nitrogen in the field for producing plant protein, 50 percent; transport of nitrogen from the farm to animal or human consumers, 75 percent. Nitrogen lost by these inefficiencies aside, it is evident that about 4.2 million metric tons of the estimated annual input are not utilized in the food chain.

It is important to realize that, without an increase in either plant or animal biomass, none of this nitrogen is really removed from the soil compartment; rather, it is merely transferred from place to place and converted from one form to another. Thus, if returns to the atmosphere are to equal inputs, 21 million metric tons must be disposed of annually. The estimates in Table 1 indicate that 1.2 million tons are in the form of human wastes, 4.2 million tons are in animal wastes, and the remaining 15.6 million tons are in other forms.

RETURN OF NITROGEN TO THE ATMOSPHERE

To determine whether nitrogen is accumulating in unwanted amounts in any compartment, the return of the 21 million metric tons total per annum to the atmosphere needs to be considered. The necessary data, however, are largely lacking. Since the annual input of 5.6 million tons in rainfall is presumed to be derived principally from ammonia and nitrogen oxides produced on earth, this figure is assigned to "returns." Stanford *et al.* (1970) estimated that 2.7 million metric tons are lost to the water compartment by erosion, and another 1.8 million tons are lost by leaching of native soil nitrogen.

Estimates of the loss of fertilizer nitrogen to waterways vary considerably. For example, leaching losses from an unusually high application of 540 kg nitrogen/ha on a sandy soil under high rainfall in Florida ranged from 0.6 to 90 kg/ha, the amount depending on the crop (Volk, 1956). Although examples of losses of nitrogen in runoff could also be cited, the early studies are of limited value when applied to modern agriculture because total nitrogen in sediments was determined. Only a fraction of this total nitrogen becomes available to plants (Barrows and Kilmer, 1963).

The available data suggest that 10–15 percent or less of the fertilizer nitrogen used annually in the United States is lost. Since about 7 million metric tons are used annually, it seems reasonable to estimate maximum losses at about 1 million metric tons. Thus, 5.5 million metric tons (2.7 by erosion, 1.8 by leaching of native soil nitrogen, and 1.0 from fertilizer) may be lost to waterways. The fate of this nitrogen is unknown, although the increased fertility attributable to nitrogen in many lakes and streams suggests that some is stored as organic nitrogen as lake sediments accumulate. Observations on some enriched lakes in Connecticut indicate that 10 percent of the nitrogen entering a lake may be retained there, a figure that is not unreasonable. Thus, about 5 million metric tons remain to be returned to the atmosphere from waterways by denitrification.

Delwiche (1970) estimated that, on a world basis, nitrogen fixation may exceed denitrification by about 10 percent. To illustrate the ranges in estimates of denitrification, losses of nitrogen by denitrification from well-drained soils have been reported to be 10–15 percent of that applied (Carter *et al.*, 1967; Dilz and Woldendorp, 1960), whereas the loss of nitrogen from poorly drained and waterlogged soils may approach 100 percent. Applying Delwiche's estimates to the United States, one finds that about 8.9 million metric tons are denitrified from the soil compartment. This amounts to a loss of about 11 kg nitrogen/ha for all land in the United States—a value that may be somewhat high. These estimates, however, make it appear that there may be a net retention in soil and water of about 1.5 million metric tons of nitrogen per annum (Table 1).

The necessary data to establish a precise nitrogen budget for the United States are largely lacking, with the exception of inputs in chemical fertilizers. Knowledge of inputs by symbiotic and nonsymbiotic fixation, in rainfall, and particularly by mineralization of soil organic nitrogen is fragmentary. Knowledge of losses to waterways and of returns to the atmosphere by volatilization and denitrification is also fragmentary. Hence, the entries in Table 1 should be considered merely as estimates, any one of which may be in error by a factor of two or more.

CHANGES IN STORAGE OF NITROGEN

Because the necessary data to establish a precise nitrogen budget for the United States are lacking and it is therefore not possible to determine whether nitrogen is being lost or gained, one might examine long-term measurements of soil or water that would indicate changes in the concentration of nitrogen stored in these compartments.

SOIL

The nitrogen content of the fertile soils of the Midwest is declining, as evidenced by the estimate that more than 3 million metric tons are released from this source every year. However, land under cultivation has decreased from a high of about 140 million ha in the early 1900's to about 117 million ha today (1972). To a large extent, this decrease of 23 million ha is accounted for by land that has been allowed to return to native vegetation. Thus, of the calculated net retention of 1.5 million metric tons of nitrogen per annum in the soil and water compartments, some is undoubtedly replenishing the native fertility of these 23 million ha. Since the inputs would consist largely of rainfall and nonsymbiotic N_2 fixation, which amount to less than 15 kg nitrogen/ha per annum, idle land could account for an annual accretion of about 0.3 million metric tons. Significant deposits of nitrate ("nitre spots") are found in the soils or geological formations in many of the western states. These natural accumulations of nitrate salts result from nitrification in these soils and represent another possible increase in stored soil nitrogen.

No evidence is available to suggest that other cropland is gaining nitrogen. It is concluded, therefore, that any increase in the nitrogen stored in soils is only modest.

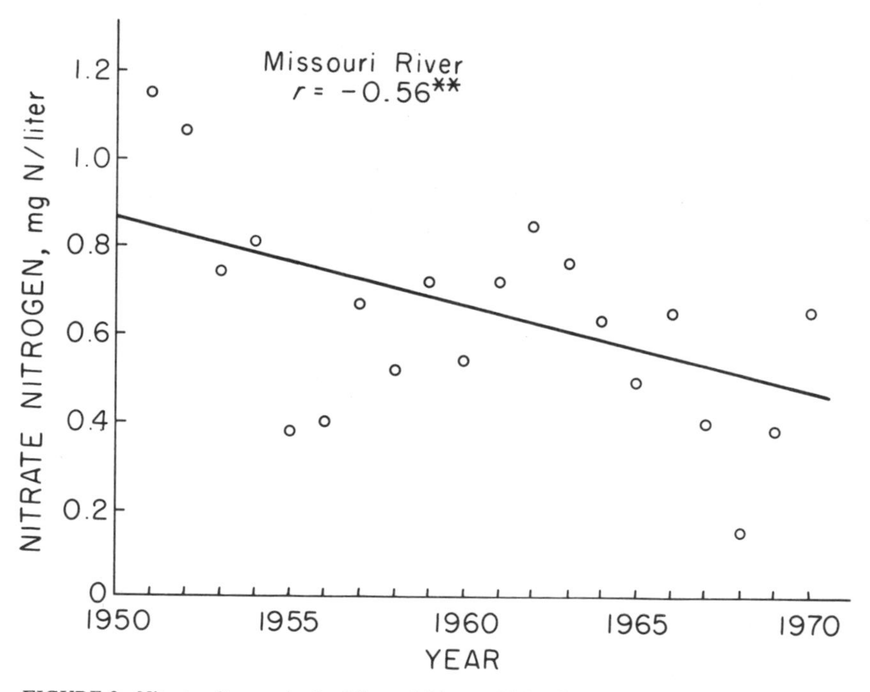

FIGURE 3 Nitrate nitrogen in the Missouri River at Nebraska City, Nebraska. (Correlation coefficient r^{**} significant at 1 percent level.) Based on data from U.S. Geological Survey.

WATER

The U.S. Geological Survey has measured concentrations of nitrogen in surface waters for many years. Its analyses of the nitrate content of six representative major rivers from 1950 to 1970 indicate a downward trend in the Missouri (Figure 3) and Brazos rivers, an upward trend in the Delaware (Figure 4), San Joaquin, and Ohio rivers, and no distinct pattern of change in the Colorado River (Figure 5). Unfortunately, the changes were not related to sources of nitrogen in the watersheds. Analyses by the Illinois State Water Survey since 1945 also indicate an upward trend in nitrate concentration in a number of rivers and streams, as well as increases in the total load carried by the streams (Harmeson *et al.*, 1971). These data are shown in Table 2 for watersheds sampled during five 5-yr periods. The average annual load is calculated from the concentrations and the related flows.

In a historical context, it is of interest to compare the values in Table 2 with figures obtained some time ago; thus, the average annual

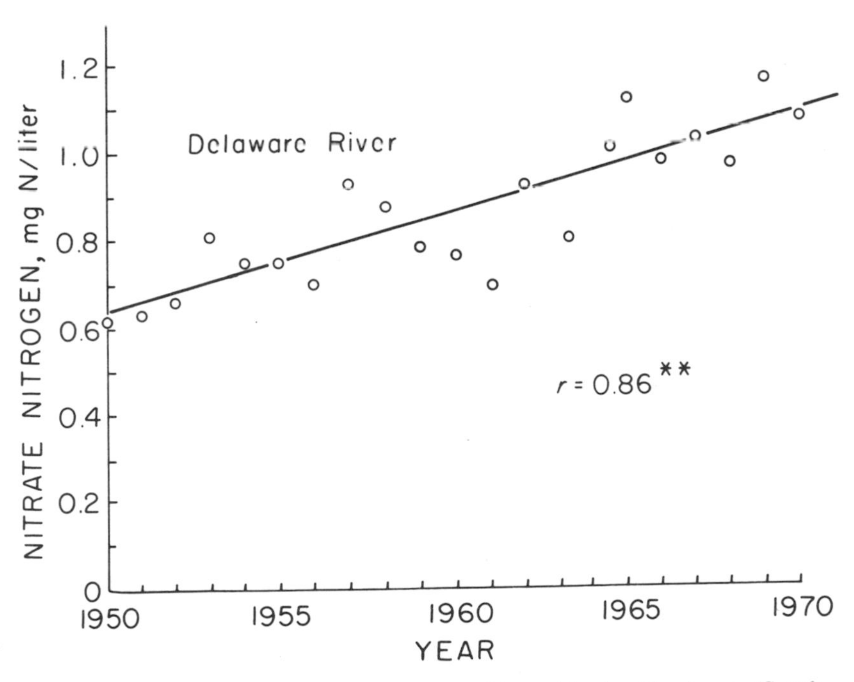

FIGURE 4 Nitrate concentration in the Delaware River at Trenton, New Jersey. (Correlation coefficient r^{**} significant at 1 percent level.) Based on data from U.S. Geological Survey.

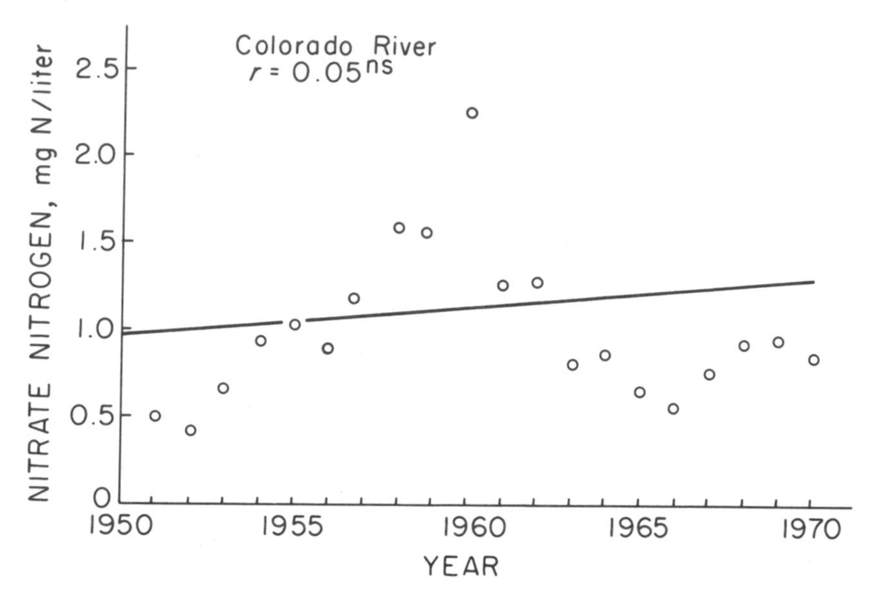

FIGURE 5 Nitrate concentration in the Colorado River at Lee's Ferry, Arizona. (Correlation coefficient *r* not significant.) Based on data from U.S. Geological Survey.

load from the Illinois River at Kampsville was calculated to be 2.33 kg nitrate nitrogen/ha/yr during 1897–1899 and 2.09 kg/ha/yr during 1900–1902 (Palmer, 1902).

A task group of the American Water Works Association (McCarty *et al.*, 1967) examined data collected since the early 1900's for a considerable number of rivers throughout the United States and concluded:

In general, the available records show NO_3-N has increased substantially since 1910 in a number of major streams in the States of Washington and Oregon, and probably some increases have occurred in streams tributary to the Mississippi River in the intensively cultivated Middle West and in irrigated regions farther west. Although increases have occurred in places, no well defined universal upward trend in the past 60 years can be seen in the available records for the eastern part of the United States. Partly this may stem from the fragmentary nature of the older records of water quality, but it also is probably partly an indication that eastern streams were already carrying large amounts of nitrogenous material before any water-quality records were obtained.

In examining nitrate concentrations in surface waters, it must be

TABLE 2 Concentrations of Nitrate Nitrogen and Calculations of Nitrogen in Runoff from Selected Watersheds in Illinois, 1945–1970[a]

	Concentration Ranges			Average Annual Load Ranges	
Period	Number of Watersheds	Median (mg/liter)	Maximum (mg/liter)	Number of Watersheds	kg/ha[b]
1945–1951	11	0.4–2.4	0.9– 9.0	10	2.3– 8.8
1951–1956	10	0.5–3.0	1.7– 6.1	10	1.1– 7.3
1956–1961	25	0.5–3.7	1.3–12.3	23	0.7–12.8
1961–1966	26	0.3–4.2	0.9–12.3	24	1.3–13.4
1966–1970	30	0.4–8.5	2.4–20.4	30	2.2–37.2

[a] Samples collected once a month.

[b] Average NO_3-N, kg/ha/yr $= \frac{\Sigma (Qi \times Ci) \times 201.7}{NA}$

where Qi = instantaneous discharge, cfs (cubic feet per second)
Ci = NO_3-N concentration, mg/liter
N = number of samples in period
A = watershed drainage area, hectares

realized that nitrate levels may be low owing to the rapid assimilation of nitrate by aquatic plants.

Although the source of this nitrate nitrogen remains a matter of controversy, it appears that nitrate concentrations have increased in some surface waters. The available data are not adequate, however, to estimate what proportion of the calculated retention shown in Table 1 might be stored in surface waters.

No long-term analyses of nitrate concentrations in groundwaters are available, so that tests of the assumption that nitrogen inputs exceed returns, thereby increasing the concentrations in soil or water, cannot be made.

CONCLUSIONS

1. Because data on the rates of biological nitrogen fixation, losses of organic nitrogen from soil, and denitrification are inadequate, a precise nitrogen budget for the United States cannot be prepared.
2. Knowledge of the extent of denitrification and of how this microbiological process can be managed in the interest of agriculture and environmental quality is seriously inadequate.
3. The amount of nitrogen in cropland is declining; the amount in land removed from cultivation may be increasing slightly.
4. Nitrate concentrations have increased in some surface waters

and decreased in others. There is little information on which to base generalizations concerning changes in nitrate concentrations in groundwater. However, nitrate concentrations in well waters in many parts of the country exceed the limits recommended by the U.S. Public Health Service for potable water supplies.

RECOMMENDATIONS

1. Monitoring of the nitrate concentration in selected surface water and groundwater supplies, to determine the extent of nitrogen losses from watersheds under different conditions of management, should be continued.

2. Data needed to establish an adequate nitrogen budget should be acquired, particularly information on the rates of nitrogen mineralization and denitrification, so that man's effect on the nitrogen cycle can be more accurately assessed.

3. Additional research on the nitrification and denitrification processes in soils at various temperatures and in various water table relationships should be undertaken.

4. Additional research on the nitrogen transformations resulting from microbial activities in streams and lakes and in soil beneath the root zone should be undertaken.

Sources of Nitrogen

Since the places where nitrogen is accumulating cannot be precisely determined by examining trends in storage, the problem of nitrogen accumulation must be assessed by asking whether there are focal points of accumulation that may produce undesired high concentrations of nitrogen in soil or water. The problem of focal points of accumulation can be dealt with by establishing the changes in concentration that may occur near the nitrogen source. For this purpose, it is convenient to divide sources into point sources–including municipal and industrial wastes, animal feedlots, septic tanks, and refuse dumps–and diffuse sources–including runoff, leaching, and tile drainage from agricultural, urban, and other land and direct contributions to lakes and streams from precipitation.

POINT SOURCES

MUNICIPAL AND INDUSTRIAL WASTES

Discharges of municipal and industrial waste water are concentrated sources of nitrogen usually released directly into surface waterways.

These discharges are frequently held responsible for undesired increases in nitrogen content of the receiving water. A well-known instance is eutrophic (nutrient-rich) Lake Erie, where about 85 percent of the total nitrogen entering the lake has been attributed to municipal wastes. In the Potomac River estuary, about 50 percent of the nitrogen appears to be derived from discharges of waste water (Jaworski and Hetling, 1970). By contrast, only 10 percent of the nitrogen reaching eutrophic Lake Mendota in Wisconsin is estimated to come from municipal and industrial wastes (Biggar and Corey, 1969).

With increasing construction of sewage treatment plants in many parts of the country, nutrients from this source will continue to increase. Thus, in many instances, discharges of waste water are likely to constitute undesirable focal points of nitrogen accumulation.

The contribution of nitrogen from domestic wastes is readily determined, since the average adult excretes essentially all the protein nitrogen consumed. Various estimates generally agree that the contribution of nitrogen from human wastes amounts to about 5.4 kg of nitrogen per person per annum. Hence, 202 million Americans would produce about 1.1 million metric tons of nitrogen per annum, an estimate that is in good agreement with that in Table 1. The most recent survey of municipal waste facilities in the United States showed that 63 percent of the population is served by sewers (U.S. Public Health Service, 1964). One may suppose that up to 75 percent are now (1972) served by sewers; thus, 75 percent of the nitrogen from human wastes is probably delivered to sewage treatment facilities. Secondary sewage treatment plants remove less than half of this nitrogen; hence, nitrogen in the effluent can be reasonably estimated at 3 or more kg of nitrogen per person per annum. The remaining nitrogen is largely incorporated in the digested sludge and presents another focal point of nitrogen accumulation.

Few data are available on nitrogen losses from sludge disposal, either from land application or incineration. Some of the nitrogen is lost by denitrification, and some may be utilized by a growing crop. Most of the remainder probably enters waterways.

Apart from nitrogen used for fertilizers, industrial consumption of nitrogen amounts to about 2.5 million metric tons, but industrial wastes contain variable amounts of nitrogen. Only small proportions of nitrogen are likely to be lost by discharge of waste water from many industrial sources, although fuel processing industries may constitute a significant source of nitrogen. Petroleum refinery wastes are very high in ammonia nitrogen, and McCarty *et al.* (1967)

concluded that these wastes "may contain quantities of nitrogen approaching that from domestic and agricultural wastes and could, therefore, be of major significance."

In view of the relatively large amounts of nitrogen consumed by industry, these sources must be examined more closely.

Food processing wastes would be expected to contain significant quantities of nitrogen. A detailed summary of the pollution potential of processing wastes has been prepared (Hoover and Jasewicz, 1967); unfortunately, it deals only with organic waste degradation in terms of biological oxygen demand (BOD), rather than nitrogen. A recent symposium on food processing wastes indicates that the nitrogen/BOD ratio of plant wastes is about 0.05 (U.S. Department of the Interior, 1970b). On the basis of a preliminary estimate, animal processing wastes are assumed to have a nitrogen/BOD ratio of 0.5. These figures are the basis for the estimates for nitrogen in food processing wastes shown in Table 3. Although these estimates are only approximations, the total of about 6.5 million metric tons of nitrogen is about 40 percent of the 15.6 million tons of "other nitrogeneous wastes" in the food chain (Table 1). Thus, the estimates are not unreasonable.

It is evident, however, that nitrogen in food processing wastes constitutes a significant point source about which little is known.

TABLE 3 Biological Oxygen Demand and Estimated Nitrogen Content of Food Processing Wastes[a]

Source	Potential Daily BOD (1,000 kg)	Estimated Nitrogen per Yr (million metric tons)
Canneries	620	0.2
Corn wet milling	60	0.02
Cotton	390	0.1
Dairy	900	2.4
Hides and leather	135	0.4
Meat	1,050	2.7
Paper and pulp[b]	16,300	0.1
Potatoes	160	0.04
Poultry	100	0.3
Sugar refining	360	0.1
Wool scouring	45	0.1
Total	–	6.46

[a]It is assumed that the nitrogen/BOD ratio is 0.05 for plant wastes and 0.5 for animal wastes.
[b]Since paper and pulp wastes have a disproportionally high BOD, the annual processing of about 110 million kg of wood (to produce 74 million kg of pulp, paper, and paper board) is assumed to be accompanied by a complete loss of nitrogen from bole wood, estimated to contain 0.1 percent of nitrogen.

FEEDLOTS

Feedlots with large numbers of animals, but in areas where there is little or no cropland on which to apply the manure, are increasingly significant point sources of nitrogen. Changes in size of cattle feedlots since 1962 are shown in Table 4. Since a steer of average weight (450 kg) excretes about 43 kg of nitrogen per annum, a 32,000-head feedlot would produce about 1,400 metric tons of nitrogen annually—an amount equivalent to the nitrogenous wastes of 260,000 people. Thus, the nitrogen pollution potential from this source is enormous. Nevertheless, little is known of nitrogen losses from feedlots. Nitrate levels under these feedlots seem to be appreciable as evidenced by the finding that nitrate nitrogen to a depth of 6.7 m in soil under feedlots in a semiarid region in Colorado averaged 1,282 kg/ha, compared with 452 under irrigated fields not in alfalfa, 233 under cultivated dry land land, 81 under native grassland, and 70 under alfalfa. Other nitrogenous compounds were also higher in water and soil under the feedlots than under adjacent soils (Stewart *et al.*, 1967). Nitrogen lost by volatilization as ammonia may be trapped by nearby water surfaces, resulting in local nutrient enrichment (Hutchinson and Viets, 1969). Where feedlots are located in more humid regions, greater runoff and downward percolation combine to enhance the possible loss of nitrogen to waterways.

Other point sources of animal manure include hog-feeding lots and poultry houses with large numbers of caged birds. These, too, are sources created by the trend toward the confinement of increasing numbers of animals on ever-diminishing areas of land. Little information is available on the extent of nitrogen losses on hog and poultry farms, although the losses are undoubtedly large under some circumstances. From the approximate rates of production of nitrogenous wastes by farm animals and the animal numbers given in Table 5, the total contribution from animal wastes can be estimated at about 6.0 million metric tons of nitrogen per year. This is somewhat higher than the estimate of 4.2 million tons shown in Table 1, possibly indicating that the efficiency of conversion of plant to animal protein is not as high as assumed in the calculations reported below ("Fertilizer and Soil Nitrogen," p. 30).

Although animal manures have been used as fertilizer for centuries, the recent trend toward confinement of large numbers of animals on small areas of land and the unfavorable economics of hauling manure to distant cropland combine to create significant new point sources of nitrogen that need considerable further study.

TABLE 4 Changes in the Number and Size of Feedlots in the United States[a]

Animals per Feedlot	Number of Feedlots								
	1962	1963	1964	1965	1966	1967	1968	1969	1970
<1,000	234[b]	231[b]	223[b]	220[b]	215[b]	210[b]	206[b]	188[b]	182[b]
1,000– 2,000	752	785	808	895	938	960	967	932	991
2,000– 4,000	373	388	421	459	486	510	522	498	543
4,000– 8,000	179	215	242	250	298	313	316	319	331
8,000–16,000	105	114	120	131	136	153	176	188	210
16,000–32,000	26	28	34	44	55	59	80	101	105
>32,000	5	7	10	8	8	13	19	31	41

[a] Source: Statistical Reporting Service (1963–1971). Some lots from the larger groups are included in smaller groups to avoid disclosing individual operations. Data are for 35 states, except for 1969–1970, for which time 12 or 13 states were excluded because of minor operations.
[b] In thousands. All others are actual numbers of feedlots.

TABLE 5 Estimated Nitrogen Production in Animal Wastes[a]

Animal	Size (kg)	Wet Manure (kg/yr)	Nitrogen (kg/yr)	Animal Population[b] (millions)	Total Nitrogen (million metric tons)
Dairy cattle	450	10,900	60	21	1.1
Beef cattle	350	7,800	43	91	2.9
Poultry	2.2	40	0.45	3,220	1.4
Swine	45	1,400	7	57	0.4
Sheep	45	680	9	20	0.2
Total	–	–	–	3,409	6.0

[a]Source: U.S. Department of Agriculture (1970).
[b]The figure for dairy cattle includes 4 million calves weighing about 90 kg, and the figure for beef cattle includes 30 million calves of the same weight.

SEPTIC TANKS

At least 25 percent of the household wastes in the United States are discharged to septic tanks or cesspools, yet little is known of the extent of losses of nitrogen from this source. Ammonium in septic tank effluent may be rapidly converted to nitrate, and this nitrate is then found at some distance from the source. Although the nitrate concentration declines with distance from the site of nitrogen discharge, its level in groundwater may exceed 10 mg nitrogen/liter near the septic tank (Preul, 1967; G. E. Smith, personal communication). Nassau County, Long Island, New York, provides a large-scale, dramatic demonstration that septic tanks can pollute groundwater. Here, the daily discharge of 300 million liters of wastes into septic tanks and cesspools has caused increases in nitrate in individual and municipal wells above the present U.S. Public Health Service drinking water limit of 10 mg nitrate nitrogen per liter (Smith and Baier, 1969).

Septic tanks, like sewage treatment plants, produce sludge that must be disposed of. Although the sludge is often trucked to sewage treatment plants for processing, it is sometimes merely hauled to unpopulated areas and dumped.

Nitrogen from septic tank effluents and sludge disposal sites constitutes a significant point source needing further study. This is particularly true in areas where septic tanks and wells are close together and where septic tanks surround water used for drinking or for recreational purposes.

REFUSE DUMPS

An average of 2.42 kg of solid wastes are collected per person per day, and more than 90 percent of these wastes goes to some 12,000 disposal sites. The nitrogen content of this solid waste is about 0.5 percent (American Chemical Society, 1969), so that about 0.8 million metric tons of nitrogen are buried in this manner yearly. A "good" sanitary landfill site is anaerobic, so that some nitrogen—at least that which is in the nitrate form before the site becomes devoid of oxygen—is lost by denitrification. Again, little is known of losses of nitrogen from dumps. It has been reported, however, that a landfill in continuous or intermittent contact with groundwater may increase the ammonium nitrogen concentration 10,000-fold over that initially present in the water. In one landfill site in Madison, Wisconsin, the losses of nitrogen to waterways were estimated to be 900 to 2,700 kg of nitrogen per annum. The loss at another landfill site in Madison, adjacent to a meat-packing waste-treatment plant, was 6,500 kg of nitrogen per annum, the landfill site contributing about 750 kg of nitrogen per annum to this total (Kaufmann, 1970). If all the refuse from Madison's 180,000 people is buried in these two sites, this would amount to about 160,000 metric tons per annum of trash containing 772,000 kg of nitrogen.

Nitrogen is apparently not lost to waterways as rapidly as it is buried in landfill sites, and these sites must therefore be considered as potential long-term sources of nitrogen.

DIFFUSE SOURCES

Nitrogen from diffuse sources, including runoff and leaching from agricultural, urban, and other land, cannot be readily identified. Hence, contributions from these sources are the subject of considerable controversy. A widely used approach to identifying sources and assessing losses is to measure runoff from watersheds and relate the observed nutrient concentrations to land use. The results to date provide some useful information. The term "runoff" in this report makes no distinction between nitrogen transport through surface runoff and transport through percolation. The nitrogen in the receiving body of water is increased by either process.

The simplest calibrated watersheds are the lysimeters located at various agricultural experiment stations. Although lysimeters, which are large tanks of soil with provisions for collecting percolating water, generally accentuate leaching and percolation (Kilmer *et al.*, 1944), they do illustrate the magnitude of losses of nitrogen that may be observed. In a review of lysimeter studies and experiments with ^{15}N-tagged fertilizers, Allison (1966) states,

> Only rarely have . . . tests shown nitrogen recoveries in the crop plus soil greater than about 95 percent of the applied nitrogen; values of only 70 to 90 percent are fairly common, and a few are as low as 60 percent. . . . Such results . . . help to explain why nitrogen recoveries in the crop under average field conditions often are no greater than 50 to 60 percent of that applied, even if immobilization is taken into account.

Although these values should not be extrapolated directly to the field, neither should one be surprised if large losses of nitrogen are observed under some circumstances.

A summary of measured annual losses of nitrogen from various watersheds is presented in Table 6. The loss of nitrogen from forested watersheds varies from 1.6 to 3.5 kg nitrogen/ha and is quite consistent in view of the wide range of soils and climates encompassed. Runoff from urban areas appears more variable. The greater variability results, in part, from varying population densities. Measurements of losses from agricultural watersheds reveal no such consistency, and the data suggest that there is no such thing as an average agricultural watershed.

The relative significance of losses of nitrogen from diffuse sources is illustrated by estimates of nitrogen from agricultural runoff in three watersheds. In Upper Klamath Lake, Oregon, 20 percent of the nitro-

TABLE 6 Estimates of Nitrogen in Runoff from Various Watersheds

Land Use	Location	Nitrogen (kg/ha)	Reference
Forested	Potomac River Basin	1.6	Jaworski and Hetling (1970)
Forested	New Hampshire	1.8	Bormann *et al.* (1968)
Forested	Ohio	2.4	Taylor *et al.* (1971)
Forested	Connecticut	3.4	Frink (1967)
Forested	Minnesota	3.5	Cooper (1969)
Urban	Potomac River Basin	2.4	Jaworski and Hetling (1970)
Urban	Cincinnati	12.9	Weibel *et al.* (1966)
Agricultural	Potomac River Basin	4.2	Jaworski and Hetling (1970)
Agricultural	Ohio	4.5	Taylor *et al.* (1971)
Agricultural	Several	0–50	Biggar and Corey (1969)

gen was estimated to be derived from agricultural runoff (Miller and Tash, 1967). In the Potomac River estuary, 31 percent was from agricultural runoff (Jaworski and Hetling, 1970), while 54 percent of the nitrogen reaching Wisconsin surface waters has been estimated to be derived from "rural" sources (Corey *et al.,* 1967). Contributions from agricultural runoff vary considerably in these three watersheds. This fact confirms earlier observations of losses of nitrogen from soil.

Thus, the nitrogen content of runoff from forested watersheds is reasonably constant. Urban runoff has been measured in a few watersheds, but it appears likely to be much more variable. Agricultural runoff is as varied as agriculture itself, and the quantity of nitrogen may range from an insignificant to an appreciable amount.

Another approach to the problem of assessing losses of nitrogen from diffuse agricultural sources involves the analysis of a variety of farming operations (Frink, 1971). Although such an analysis may not always reveal the exact magnitude of the losses, it may disclose ways in which the losses may be minimized. For example, rather than monitoring runoff from the 26 million ha where corn is raised in the United States, comparative losses of nitrogen when corn is grown by different farm practices can be assessed. Where this approach has been used, it has been noted that timing of fertilizer application has a great effect on the efficiency of utilization of nitrogen fertilizer by corn (Lathwell *et al.*, 1970). Furthermore, comparisons of the results of certain agricultural operations in irrigated areas in California have suggested that some crops can remove nitrate from groundwater if fertilization is adjusted to crop need (Stout and Burau, 1967). In a study of three irrigated areas along the upper Rio Grande, where fertilizer use has increased as much as 100-fold during 1934–1963, no measurable increase in the nitrate content in irrigation return drains to the river was found. The apparent removal of nitrogen from the water may be the result of denitrification of leached nitrate in the anaerobic zone near the water table (Bower and Wilcox, 1969). Similarly, dairy farm operations in the Northeast were analyzed as factories for the conversion of inputs in feed and fertilizer into outputs in milk and meat (Frink, 1969, 1970), and it was concluded that losses of nitrogen per unit of land area increased considerably as available cropland decreased, reaching 225 kg nitrogen/ha when only 0.4 ha was available per cow. Thus, losses of nitrogen from manure seem enhanced under intensive land use in the humid Northeast, a situation probably quite different from that prevailing in feedlots of the arid Midwest.

Man, through his technological achievements, is able to make ammonia from atmospheric nitrogen, and nitrate from ammonia. These man-produced forms of nitrogen are exactly the same as those produced by natural biological processes. The farmer uses industrially produced nitrogen fertilizers to supplement the diminishing levels of available nitrogen in the soil. He can apply them to his soil directly, without having to wait on soil microorganisms to supply the equivalent amount of nitrogen for his growing crop. He also often finds that applying fertilizer nitrogen is much less costly than trying to acquire additional fixed nitrogen by growing legumes, which have the unique ability to use atmospheric nitrogen. Total cost differentials can be so great as to mean the difference between continuing business as a farmer or facing bankruptcy.

A significant source of fixed nitrogen is the internal combustion engine. Nitrogen oxides emitted with engine exhausts in the United States amount to several million tons of fixed nitrogen per annum. Another 6 million tons of fixed nitrogen are released as coal is burned. Some 400 million tons, containing about 1.5 percent of nitrogen, were burned in 1968 (Gallagher and Westerstrom, 1970). When coal is burned, nitrogen leaves the smokestack as nitrogen oxides, eventually to be returned to the earth's surface as nitrate. Thus, the nitrogen oxides produced by automobiles and by coal burning contribute significant amounts of nitrogen to the atmosphere, possibly enough to be an important nitrogen source for forest lands.

In addition to fixed nitrogen dispersed from coal burning and engine exhausts, industrial nitrogen fixation in the United States will amount to about 11 million metric tons in 1971. About 2.5 million tons will go into domestic steel, rubber, plastics, and other industrial uses, and about 7.5 million tons will go into domestic fertilizers. The remainder will probably be exported. There is considerable uncertainty as to how much of the fixed nitrogen used in domestic industries is recycled into the general environment, but all these sources, totaling about 20 million tons per annum, are eventually dispersed to the biosphere by a variety of routings, some of which have not been adequately traced.

NITRILOTRIACETIC ACID

Nitrilotriacetic acid (NTA) is a chemical designed for use in detergent formulations. When introduction of NTA into this nation's waterways began, about 1970, it was anticipated that enormous quantities would be introduced. Government action resulted in a voluntary termination of the use of this compound in the United States for detergent formulations, pending the development of fuller information concerning its effects on man and waterways. The NTA episode illustrates the folly and potential danger of our present policy, which does not require pretesting of substances that can be introduced on a large scale into waters. Inasmuch as NTA contains nitrogen, once the compound is degraded by aquatic microorganisms, it would serve as a nutrient contributing—probably to a small extent—to the development of algal blooms. As a chelating agent, it brings metals into solution or maintains them there. As a tertiary amine, it could decompose to secondary amines, which, if conditions allowed for their accumulation in water in the presence of nitrite, might react to yield the highly hazardous nitrosamine class of compounds. The postulated dangers may not be real; indeed, it appears that most of them were never subjected to close scrutiny. Nevertheless, industry was allowed to begin production and widespread dissemination of a chemical whose use might be objectionable for environmental and public health reasons. Should NTA indeed have untoward ecological or health effects, we are fortunate in having its production abandoned. On the other hand, should NTA be innocuous, society has been deprived of a useful substitute for the phosphate in detergents, and industry has been deprived of a legitimate market.

Many nitrogenous chemicals that enter our streams, rivers, and lakes are of value in the home, on the farm, or in industry, but there are few feasible ways to keep them out of the waterways. Some, like NTA, may be designed to replace a substance that has an undesirable effect in water, on land, or in some domestic or industrial activity. A few may cause changes in the fish population, affect the quality of water, or be hazardous to human health. In view of the continued entry of many nitrogenous chemicals into bodies of water and the potential danger of a few, it is essential that public agencies decide—before the compounds are marketed—which widely used chemicals can and which should not be introduced, directly or indirectly, into our waters.

CONCLUSIONS

1. Runoff from agricultural land varies in amount, and the nitrogen entering waterways from such land may range from an insignificant to a major amount. Little attention has been paid to nitrogen

contributions from urban runoff. Contributions of nitrogen to waterways from forest land seem to be uniform. Nitrogen from all these sources can be adequate for the growth of aquatic plants.

2. Discharges of municipal waste water are often a significant source of nitrogen for waterways. Appreciable contributions of nitrogen may come from septic tanks and sludge-disposal sites, particularly where septic tanks are located near wells or surface water supplies.

3. The recent trend toward feeding large numbers of animals on small parcels of land and the unfavorable economics of hauling manure to distant cropland combine to make feedlots significant point sources of nitrogen.

4. Little is known of the contribution of industrial and food-processing wastes to nitrate levels in water. Refuse dumps are probably long-term sources of nitrogen.

5. Large quantities of nitrogenous chemicals—including nitrilotriacetic acid (NTA), proposed for use in detergents—can be introduced into waterways, but there is no governmental requirement that information on their behavior and effects be obtained before they are released.

RECOMMENDATIONS

1. Additional effort should be directed toward the control of nitrogen entering waterways from both diffuse and point sources.

2. Where nitrogenous agricultural wastes are significant problems, local control should be exercised and action programs developed. This can be on a river basin or other area basis with appropriate monitoring by state and federal agencies.

3. To minimize the introduction of nitrogen into water, alternatives to sanitary landfill, such as composting, and means for recycling human and animal wastes should be developed.

4. The present Public Health Service recommended limits for nitrate in drinking water should not be relaxed.

5. The distribution in the environment of nitrogen-14 and nitrogen-15 should be determined. The availability of low-cost stable isotopes of nitrogen would be extremely valuable for research on nitrogen transformation in soils and water.

6. The extent of nitrogen movement from point sources, especially animal feedlots, into water should be determined.

7. The movement and persistence of nitrogenous chemicals, the products of biological and nonbiological degradation of these chemicals, and the potential effect of the chemicals on man and aquatic life should be established before nitrogenous chemicals are allowed to be introduced on a large scale into waterways.

Fertilizer and Soil Nitrogen

FERTILIZER USE AND DIETARY PATTERNS

Rapidly expanding use of nitrogen fertilizers in agriculture, particularly in Europe and North America, has raised questions about the implications of this trend for environmental quality and public health. Attention is focused on nitrate because of its presence in aerated soils, its movement through and over soil and into water, and its tendency to accumulate in certain plants consumed by man and livestock.

Changes in agriculture seldom occur rapidly; when they do, it is usually for compelling reasons. It is, therefore, necessary to assess the reasons for modern trends toward increasing use of nitrogen fertilizer in U.S. agriculture. At the same time, it is essential to determine, insofar as possible, the degree of environmental hazard or disturbance that might result from expanding use of fertilizer and to consider the benefits accruing to consumers of agricultural products.

The discussion that follows is concerned with population, food consumption, and agricultural nitrogen from 1910 to 1968.*

*Unless otherwise indicated, data are from U.S. Department of Agriculture (1963, 1969).

Agriculture's principal function is to provide food and fiber for man's use. In this country, demands for quantity, quality, and variety in food are determined in the market place, where price and quality adjustments are finally made. The food market thus influences the quantities of nitrogen used on the farm where the food originates.

Consumer choices in protein-food are predominant among the factors determining the amount of agricultural nitrogen required at the food-producing farm sites. For this reason, changes in the amount of nitrogen used in farming will lead to changes in the kinds and amounts of protein-food that can be made available to consumers. This is true in any country that depends mostly on indigenous agricultural resources for its food supply.

The American people demand a high proportion of protein, which is obtained in foods such as meats, poultry, eggs, and dairy products. Table 7 gives the 1968 consumption pattern of major agricultural products in this country and the amounts of protein and protein nitrogen derived from each category. The calculations show that Americans consume about 70 g of animal protein and about 29 g of vegetable protein per capita each day, which is equivalent to about 16 g of nitrogen per person per day.

A sufficient amount of nitrogen has to be made available in plant-usable forms at American farm sites each year to provide the nitrogen component of these protein foods. This farm-site nitrogen requirement is met, in part, by mineralization of organic nitrogen in the soil, by such nitrogenous substances as ammonia and nitrate in rainwater, by fixation of atmospheric nitrogen by microorganisms, and by the use of organic and inorganic nitrogenous fertilizers.

It is evident from the calculations in Table 8 that the protein nitrogen demanded from our food production system is greatly increased because of the amount of plant protein used in feeding animals. Thus, while each person in the United States directly consumes an average of 3.7 lb of nitrogen from plant protein each year, about 61 lb of plant protein nitrogen per person per year are fed to the domestic animals that provide the animal protein in the meat, milk, eggs, etc., being consumed by Americans.

Consideration must also be given to the efficiency of nitrogen use on the farm. Much of the fertilizer nitrogen applied to agricultural soils does not appear in our food for two main reasons: Crop recovery of applied nitrogen is about 50 percent; and product losses are high between the time of seed planting and ingestion of the resulting food

TABLE 7 Per Capita Consumption of Protein-Containing Foods, Protein, and Protein Nitrogen in the United States, 1968

Source	Pounds per Capita per Yr[a]	Estimated Average Percent Protein	Protein per Yr (lb)	Protein per Day (g)	Nitrogen per Day (g)
Animal sources					
Meats (all)[b]	182.7	16	29.4	36.6	5.86
Beef	109.4	–	–	–	–
Veal	3.6	–	–	–	–
Lamb and mutton	3.7	–	–	–	–
Pork (lean)	66.0	–	–	–	–
Poultry (all)	45.0	16	7.1	8.8	1.41
Chicken (ready to cook)	37.1	–	–	–	–
Turkey (ready to cook)	7.9	–	–	–	–
Fish[b]	11.0	18	2.0	2.5	0.40
Eggs	45.0	12	5.4	6.7	1.07
Whole milk	279.0	3.3	9.2	11.4	1.83
Cheese	11.0	20	2.2	2.7	0.43
Ice cream, etc.	27	5	1.4	1.7	0.27
Subtotal	–	–	56.7	70.4	11.1
Plant sources					
Fruits (fresh)	78	1	0.8	1.0	0.16
Fruit (canned, frozen)	44	1	0.4	0.5	0.08
Vegetables (fresh)	97	2	1.9	2.4	0.38
Melons	22	1.2	0.3	0.4	0.07
Potatoes	112	2.8	3.4	4.2	0.67
Wheat flour	112	11	12.3	15.3	2.46
Dry beans, peanuts	12	22	2.6	3.2	0.51
Corn, rice, etc.	16	9	1.4	1.7	0.27
Subtotal	–	–	23.1	28.7	4.60
Total	–	–	80	99	15.9

[a] Source: Newspaper Enterprise Association (1969).
[b] Fish and wild game are excluded from calculations pertaining to agricultural nitrogen requirements at the farm site.

by the human consumer. A product loss of about 25 percent is ascribable to combinations of drought, hail, fire, insects, plant pathogens, acres planted but not harvested, death of animals, food processing wastes, kitchen wastes, spoilage in storage, and household spoilage. Part of the nitrogen added on the farm is not consumed because kitchen wastes and household spoilage result in a discarding of the products. Therefore, possibly only three eighths of the plant-available nitrogen mobilized in or added to soils on the farm finds its way into the plant protein actually eaten by the American consumer. Much nitrogen is, of course, lost in the conversions of plant protein to ani-

TABLE 8 Sources of Human Dietary Protein in the United States and Nitrogen Used or Consumed per Capita for Each Protein Source, 1968

Protein Source	Protein Consumed as Human Food (g/person/day)	Plant Protein to Animal Protein Conversion Factor	Plant Protein Used or Consumed (g/person/day)	Nitrogen in Plant Protein Used or Consumed	
				lb/person/yr	kg/person/yr
Animal					
Meats	36.6	9	330	42.5	19.4
Poultry	8.8	6	52	6.7	3.0
Fish	2.5	–	–	–	–
Milk	11.4	4	45.7	5.9	2.7
Cheese	2.7	4	10.8	1.4	0.63
Eggs	6.7	4	26.8	3.5	1.57
Ice cream, etc.	1.7	4	6.8	0.9	0.40
Subtotal	70.4	–	473	60.8[a]	27.7[a]
Plants					
Cereals	17.0	–	17.0	2.20	1.00
Fruits and vegetables	8.5	–	8.5	1.10	0.50
Legumes	3.2	–	3.2	0.41	0.18
Subtotal	28.7	–	28.7	3.71[b]	1.68[b]
Total	99.1	–	502	64.5	29.4

[a] Nitrogen contained in the vegetable protein before conversion into animal protein consumed as human food.
[b] Nitrogen contained in vegetable protein consumed directly as human food. Nitrogen is taken as 0.16 the weight of protein.

mal protein. Taking into account these losses, only about one fourteenth of the nitrogen made available to crops is consumed in human food.

Estimates of the nitrogen required each year to provide the food and fiber that reach the American consumer are shown in Table 9. A surprisingly large amount, 15.1 million metric tons of farm-site nitrogen, goes toward providing animal protein, whereas about 0.94 million ton provides Americans with their consumable plant protein. Sugar and cotton together require about 0.77 million ton of farm-site nitrogen.

Although the data in Tables 7, 8, and 9 will be subject to adjustments and refinements as additional information becomes available, it is clear that the overwhelming preponderance of nitrogen used on our farms is associated with American dietary preferences for meats and nonmeat proteins of animal origin.

A graphic account of the changing American agricultural and food consumption scene from 1910 to 1968 is given in Figures 6 and 7.* During this time, the population grew from 92 million to 202 million, draft animals disappeared, cultivated acreage decreased, and substantial changes in dietary patterns took place. In 1968, as in 1910, an individual consumed about 100 g of protein per day, but the protein sources changed materially over that period. Per capita consumption of fish and milk products changed little throughout this period. In contrast, the per capita consumption of flour, cereals, and potatoes declined steadily, and consumption of eggs, poultry, and meat increased (Figure 6). The small dots in Figure 6 represent the actual data for annual per capita consumption; the smooth curves show the trends. The consumption of animal products (eggs, poultry, and meats) held steady or declined slightly between 1910 and 1930 and then began a sharp upward trend that still persists.

The results of an increasing U.S. population combined with a growing preference for high-protein foods of animal origin are shown by the total of meat, poultry, and eggs consumed (Figure 7). There appears to be no leveling off of this sharply rising curve of animal-protein foods consumed even though the population growth rate appears to have declined between 1960 and 1968. The two curves of Figure 7, showing rising use of nitrogen on the farm and the total of meat, poultry, and eggs consumed, are strikingly similar.

*Complete data for 1969–1970 were not available when this report was prepared.

TABLE 9 Major Agricultural Nitrogen Requirements for Food and Fiber Consumed in the United States[a]

	Protein Consumed (g/person/day)	Product Consumed (lb/person/yr)	Nitrogen at 37.5% Use Efficiency: Required Yearly per Person (lb)	Nitrogen at 37.5% Use Efficiency: Required Yearly for Population of 202 million (million metric tons)
Edible protein from				
Animals (fish excluded)	67.9	54.6	163.8	15.1
Plants	28.7	23.0	10.2	0.94
Sugar (refined)				
Beet and cane	–	98.9	6.6	0.61
Fibers				
Cotton	–	22.5	1.8	0.16
Wool	0.1	1.0	–	–
Wood products				
Lumber	–	840	–	–
Paper products	–	436	–	–
Total	–	–	182.0	16.8

[a] The calculations assume 50 percent utilization of nitrogen by plants at the farm and 25 percent product loss from all causes (pet animals and miscellaneous small crops excluded). The 25 percent product loss includes maintenance of sire and dam of meat animals, unharvested plantings, wind and storm damage, death losses, cullings, trimmings, peelings, condemnation rejects fire, storage losses, and all other product or maintenance losses that have drawn on farm nitrogen. Animal protein (exclusive of fat and moisture) includes protein from beef, pork, veal, mutton and lamb, chicken, turkey, eggs, milk, ice cream, and cheese. Fish have been excluded as being nondomesticated and not demanding of N fertilizer; however, Americans consume 2.0 g of fish protein/person/day. Plant protein (exclusive of fat, carbohydrates, and moisture) includes protein from cereals, potatoes, vegetables, rice, wheat flour, fruits, and legumes. Sugarcane fertilizer nitrogen requirements are assumed to be equivalent to the nitrogen required by sugar beets, namely, 1 lb of nitrogen for 15 lb of refined sugar.

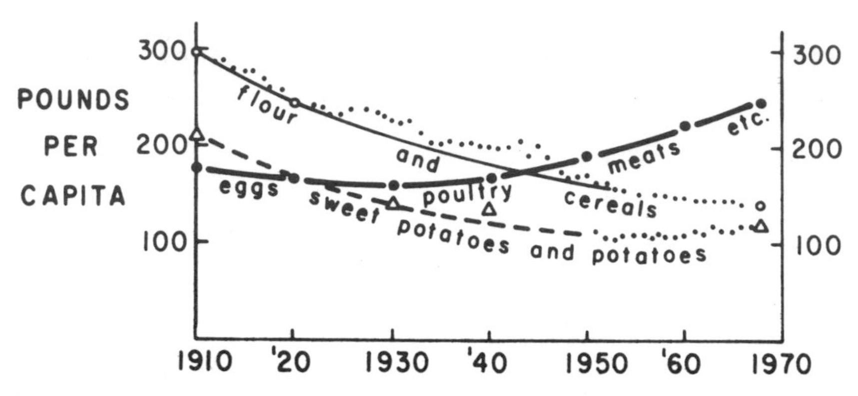

FIGURE 6 Changing patterns in consumption of certain foods in the United States, 1910–1968. Source of data: U.S. Department of Agriculture (1963, 1969).

From the foregoing, it is evident that the amount of plant-available nitrogen needed in soils that support a population on a vegetarian diet is much less than the amount needed in soils that support a population on the usual American nonvegetarian diet. For example, a predominantly vegetarian diet with small amounts of milk and meat can be had for as little as 23 lb of plant-available nitrogen per capita per year; in contrast, 179 lb of farm-site plant-available nitrogen per capita per year were required to support the 1968 American diet. However, if animal protein consumption continues to rise, use of fertilizer nitrogen will similarly increase. Although dietary patterns might be changed toward substituting vegetable protein for some of the animal protein now being consumed, thereby reducing fertilizer nitrogen requirements, such a change will depend less on farmers and agricultural scientists and more on experts in human nutrition and the influence that they may have in guiding food choices by the American consumer.

As is evident in Figure 8, a marked upswing in the consumption of meat, poultry, and fish groups began in the 1940's. The upper curve is a plot of year-to-year data for pounds per capita of meats, poultry, and fish available in the United States between 1910 and 1968. (The smooth curve joins points of the census years.) The lower curve shows the grams of protein available per capita per day. After the downward trends reflected in both curves for the period before 1930, the trends changed direction as opportunities became available for expansion of the nitrogen industry.

The surprising aspect of the 1940–1955 period was the remarkable speed with which agricultural research and technological develop-

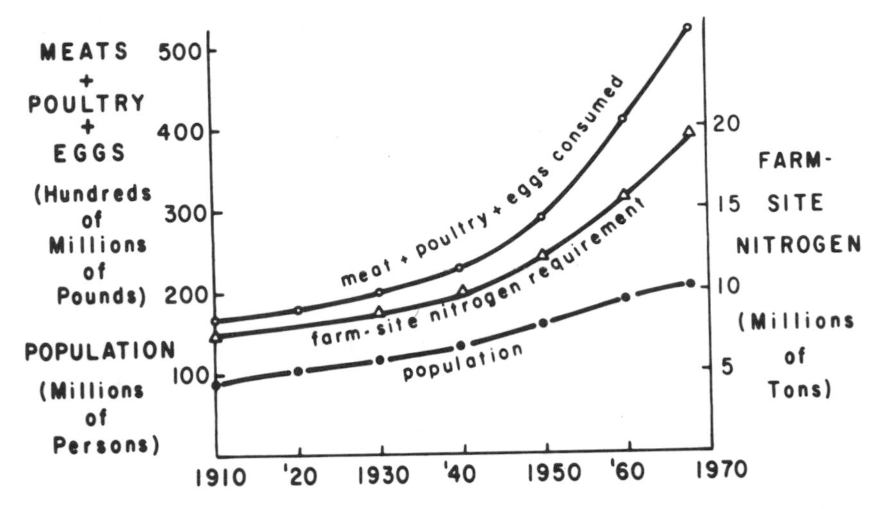

FIGURE 7 Trends in consumption of high-protein foods and in farm-site nitrogen requirements in relation to the U.S. population, 1910–1968. Source of data: U.S. Department of Agriculture (1963, 1969).

ments in mechanization and industrial nitrogen fixation were able to fill the growing need for nitrogen. Major contributions to the upturn in the amounts of nitrogen that could be made available for, saved for, or diverted to food production were progressive reductions in the acreage required for feeding the horse and mule population; improved legume pastures and hay crops for other livestock; a particularly successful repayment from soybean research that resulted in more than 41 million acres of soybean plantings by 1968; considerable improvements in the vegetable protein to animal protein conversion efficiencies of dairy cattle, poultry, swine, and beef animals; rapid adoption by farmers of soil-conserving practices; and new varieties of cereal grains that responded to nitrogen fertilization with greatly enhanced yields.

The significance of higher yields from fewer acres is illustrated by land use for maize and wheat. The 1968 maize crop was harvested from 55.7 million acres yielding 78.4 bushels per acre, whereas the 1940 crop came from 86.4 million acres yielding 28.4 bushels per acre. If nitrogen fertilizers were removed from the present scene, productivity would gradually decline and would eventually revert to the 1940 yields and even lower. Reserves of native soil nitrogen would be further depleted. If the productivity of American farms should return to 1940 levels, 98 million additional harvested acres

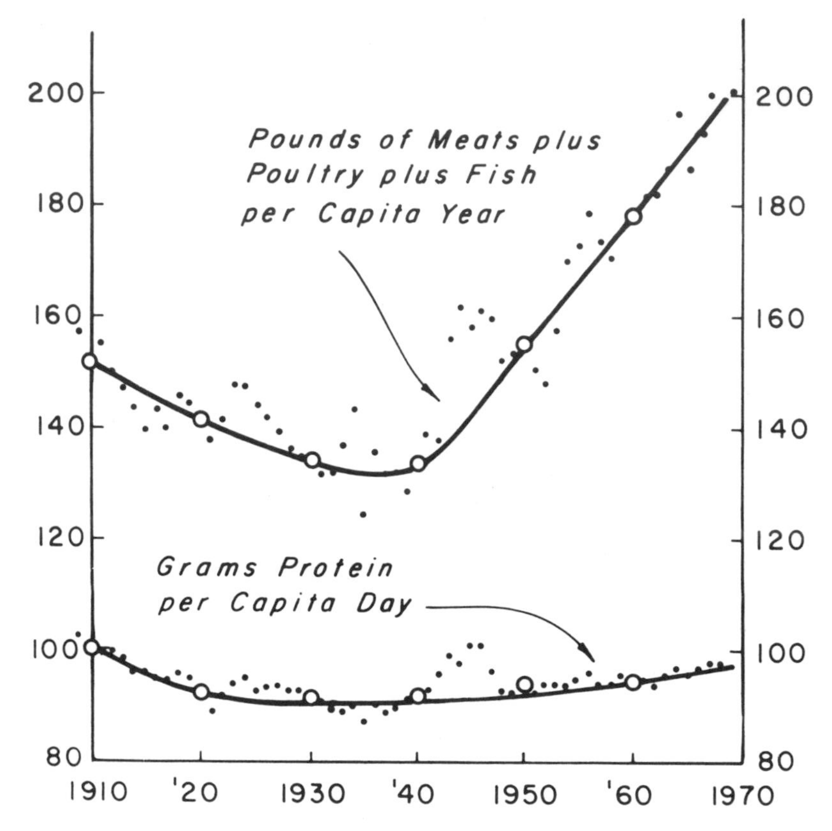

FIGURE 8 Trends in per capita consumption of meat, fish, and poultry in relation to total protein intake in the United States, 1910–1968. Source of data: U.S. Department of Agriculture (1963, 1969).

of corn would be required to produce the equivalent of the 1968 crop. Similarly, the 1968 wheat yield per acre would fall from 28.4 to 15.3 bushels and an additional 50 million harvested acres of wheat would be required to produce the equivalent of the 1968 crop. The country does not have 148 million acres of new cropland unless we gamble once more with erodible soils.

AGRICULTURAL NITROGEN AND SOIL NITRATE

Agricultural processes affect the uptake of nitrate into plants and the rate at which nitrate is formed in soil. The farmer controls the time of planting and harvest, and this man-governed timing causes nitrate to appear in the soil solution as pulses associated with tillage, manuring, and fertilization.

Plants extract both water and soluble nutrients, including nitrate, from the soil solution. Excess water passing through soils is known as the soil leachate. Leaching is essential in removing a variety of salts from soils, and it is in this manner that nitrate derived from the decomposition of plants and soil organic matter is removed.

Natural pathways of water and incidental salt movement have been altered by man in many important ways in the interest of flood control, water conservation and storage, land reclamation, and irrigation. As water is stored and rerouted, old problems in water management are solved and new problems arise. Among these new problems, the disposition of salts from soil leachates is perhaps the most difficult. Whenever new lands are brought under cultivation by draining waterlogged areas or by irrigating drylands, new sets of biological and hydrological activities prevail. For example, when land is reclaimed, either from swampy areas or from the sea, as with the polders of the Netherlands, high-quality surface water is diverted through the soil and thence to drain tiles and ditches leading to disposal sites for the soil leachates. If such leachates are discharged into streams, the streams acquire additional burdens of salt, including nitrate, that were not important components of the hydrological system before land reclamation was undertaken.

In their review of factors affecting the accumulation of nitrate in soil, water, and plants, Viets and Hageman (1971) stated:

> Growth of all organisms requires nitrogen. The photosynthetic organisms, whether they be crops on land or phytoplankton and algae in water, require nitrogen either as ammonium or nitrate, although they may use some simple organic nitrogen compounds. Hence, ammonium, total soluble nitrogen, and nitrate cannot be considered independently.

Improved fertility nearly always means increased soil nitrogen levels and greater potentials for nitrate to escape in soil leachates. The amount of nitrate escaping may vary considerably, however. It may be essentially nil if subsoils are anaerobic, or it may be very high if the soils are well aerated and heavily fertilized. Thus, Viets (1971) summarized the nitrate nitrogen content of water tables (in parts per million as follows: water tables underlying virgin grasslands, 0.1–19; those underlying fallowed wheat, 5–9.5; those underlying irrigated crops (except alfalfa), 0.36; those underlying irrigated alfalfa, 1–44; and those underlying cattle feedlots, 0–41. The range is frequently very broad, even within a limited geographical area, such as the one in northeastern Colorado from which the foregoing data were obtained. Moreover, because of microbial nitrification and denitrification in soil, as well as nitrate absorption by plant roots, no simple rules or criteria can be set forth to cover all cases. The rapid rise and fall of nitrate levels is illustrated by the short period, a single week, required for the drainfield of a sewage-treatment plant to build up the nitrate-nitrogen concentration in soil solutions from nil to several hundred parts per million and for the nitrate-nitrogen concentration to drop to nil again when the weekly sewer-plant discharge into the drainfield is resumed (Stout and Burau, 1967). Another difficulty in establishing meaningful generalizations or rules is that applications of nitrogen fertilizer to crops sometimes reduce nitrate leaching losses, as compared with fallowing to release native soil nitrogen; thus, Allison *et al.* (1959) showed that fertilization with cropping resulted in much less nitrate loss than fallowing during a 5-yr study period.

The runoff and drainage characteristics of a watershed are strongly influenced by topography, plant cover, temperature, season, and the intensity and duration of rains. For example, in areas where soils are frozen during the winter, management problems differ considerably from those of frost-free areas. Differences in climate and moisture induce changes in nitrate concentrations in soil, affect the rates of nitrification and denitrification, and influence the amount of nitrate that might be lost with runoff or by leaching. Because of the number of these variables, and because the quantitative influence of many of the variables cannot be predicted accurately, it is important that any regulations that might be promulgated as a result of nitrogen requirements for crop production be based on individual watersheds, local climate, soil types and their relationship to topography, crops, cropping patterns, and degrees to which agricultural activities might give rise to degradation of water supplies.

Nevertheless, it is essential to determine whether industrially supplied nitrogen fertilizers have become so cheap that they are being used too liberally, to the detriment of the environment. The determination is made difficult by a number of built-in constraints and penalties on overuse of nitrogen fertilizers. Lodging of wheat and rice crops illustrates one type of constraint. Even modest applications of nitrogen may cause enough stem elongation and weakening to result in breakage of the stems under the weight of the heads of grain. The problem does not affect hybrid maize or the new short-stemmed wheat and rice varieties being introduced into some of the developing countries. Sugar production from beets illustrates another type of constraint. To increase the sugar content of the beet root, sugar beet plants must be reduced to a stage of nitrogen deficiency during the latter part of the growing season; if the nitrogen supply is allowed to remain high, products of photosynthesis combine with nitrogen to promote the formation of leaves, rather than be stored as sugar. In grain crops, excessive nitrogen may produce an early spring flush of vegetative growth that uses up available soil moisture before heads are set, thus reducing the yield of grain. These kinds of management constraints on overuse of nitrogen fertilizers must be kept in mind in assessing the significance of recent trends toward increasing use of fertilizer nitrogen.

Still, there are instances of excessive use of nitrogen fertilizers, and individual farmers or even extensive areas may be the sources of nitrate that reaches water supplies in undesirable concentrations. Misuse must be evaluated and controlled, but it is necessary to bear in mind the need for using fertilizers properly and to avoid condemning all agricultural operations that rely on nitrogen fertilizers in a given area.

Foliar application might provide a very useful alternative to applying to the soil all the fertilizer nitrogen needed by certain crops. A greater efficiency in nitrogen uptake has been demonstrated to occur as a result of foliar application (Mukherjee and De, 1968; Wittwer and Teubner, 1959), and there would be less nitrate buildup in soil and minimized leaching losses. Many crops respond to foliar application, including cereal grains, leafy vegetables, and forage plants.

The lack of direct means for identifying the source of the nitrogen in many groundwaters and surface waters adds to the difficulty of evaluating the extent to which increasing nitrate levels in a watershed are due to fertilizers, and it complicates the task of locating farmers who may be applying excessive amounts of nitrogen. It is evident, however, that if the conversion of farm-site nitrogen to plant protein is only 50 percent efficient, the potential for loss of nitrogen from soil and its entry into water is considerable—whether the nitrogen source is fertilizer, recycled animal manures, mineralized organic nitrogen of soil, nonsymbiotically fixed nitrogen, or nitrogen deposited with rainfall. Clearly, more research and development effort should be directed toward improving the efficiency of conversion of nitrogen to plant and animal protein.

The use of nitrogen in American agriculture is likely to increase in the future, and there is no evidence for a diminution in the rate of rise of the farm-site nitrogen requirement. Therefore, environmental problems related to nitrate will undoubtedly be intensified. On the other hand, if the population growth and the trend toward more animal protein in American diets should cease and if exports should be discontinued or reduced, the farm-site nitrogen requirement curve would tend to level off. The need for nitrogen on the farm might even decline if research were to greatly improve conversion of vegetable protein to animal protein or efficiencies of nitrogen use in crop production. If this nation's population continues to grow and if demands for higher proportions of animal protein in the diet continue to rise, as is quite likely, it is especially important to accelerate research activities designed to increase efficiency of nitrogen use in both plant and animal production.

It is possible that economical methods for removing nitrate from drinking waters will become feasible and acceptable. Drinking waters constitute only a small fraction of the total water that is used for domestic purposes, and it is conceivable that at some time in the future nitrate will have to be removed from supplies of drinking water, despite attempts to prevent loss of nitrate from the land. The water softener is a common household accessory that removes calcium, iron, and magnesium. Nitrate could be removed as easily, but at a significant cost.

Certain activities on the farm frequently pollute bodies of water. The unrestricted drainage of barnyards into streams was at one time not uncommon. The decomposing remains of animal manure spread on frozen ground are subject to washing into streams with runoff from spring thaws. Considerable progress has been made in recognizing the problem, if not in applying methods for reducing pollution from such sources. Means for minimizing nitrate discharge associated with the decomposition of animal wastes include the use of agronomic practices that conserve nitrogen applied in the field, adding manure to the soil at levels that will ensure that much of the nitrogen will be assimilated by crops grown on that soil, and promoting the microbial reduction of nitrate to gaseous nitrogen.

THE INFLUENCE OF FERTILIZERS ON SOIL ORGANISMS

It is often stated that chemical fertilizers are detrimental to organisms and biological processes in soil that are essential for plant life and crop production. In this country and abroad, soil scientists and microbiologists have investigated the possibility that nitrogenous fertilizers are harmful to the microbiological processes on which plants rely. It is clear from many experiments that nitrogen fertilizers at the rates used in agriculture do no long-term harm to these processes.

Anhydrous ammonia fertilizers may inhibit certain of the microscopic inhabitants of the soil, but the effect is localized and transitory. Nitrogen from fertilizers also inhibits the activity of microorganisms that assimilate atmospheric nitrogen and convert it to forms ultimately used by plants. However, the plants are not harmed, because they still receive the nitrogen. Moreover, when the available nitrogen content of these soils declines, these beneficial microorganisms again are able to use atmospheric nitrogen. In addition, when nitrogen from either industrial or natural sources is oxidized by certain specialized bacteria, the acidity of the soil may increase. This acidification is a function of the amount of nitrogen. Whether it is derived from a natural or synthetic process is unimportant. Pronounced acidification is unquestionably injurious to many plant species and to various biological reactions in the soil, but it is readily countered by liming.

CONCLUSIONS

1. The rapid increase in use of nitrogen fertilizer in the United States in recent years has been caused by increases in quantities of food required by a growing population, increases in consumer demands for animal protein in lieu of vegetable protein, and continuing depletion of the reservoir of native pretillage organic nitrogen in soils.
2. Large quantities of nitrogen must be obtained by plants from the soil to provide adequate food for the American population; in the process, soil nitrogen is converted to nitrate—the form of this element most readily absorbed by plant roots. Whether the nitrogen comes from that which is naturally present in soil or is derived from commercial nitrogen fertilizers, the probability of nitrate escape with leaching waters increases with intensified agricultural operations.
3. Efforts to obtain higher yields per unit of land area through fertilization, whether the fertilizer is inorganic or organic, nearly always mean greater potentials for nitrate to be carried into waterways. Moreover, the percentage of applied nitrogen recovered by the

crop decreases as the rate of nitrogen application increases. When the efficiency of nitrogen use becomes low, greater losses of nitrogen occur, particularly from well-drained soils, and the nitrogen may then escape to leaching waters and thus to streams.

4. Increments of increasing fertilizer applications provide diminishing increments in crop returns and contribute to greater losses of nitrogen to bodies of water. The low cost of fertilizers tends to encourage inefficient use of nitrogen fertilizer on the farm.

5. Because intensification of agriculture favors greater decomposition of native soil nitrogen, often involves greater use of inorganic nitrogenous fertilizers or animal manures, and frequently leads to greater activity by certain groups of microorganisms in soil, the nitrate content of underground waters may rise. However, in oxygen-deficient soils with high water tables, the nitrate may be converted to nitrogen gas.

6. In humid regions, the nitrate concentration in water percolating through cultivated soils is a function of the fertility level of the soils. The amounts of water percolating through the soil at any given time, the degree of nitrate removal by crops, and the activity of denitrifying microorganisms determine the nitrate concentration in soil leachates.

7. Industrial nitrogen fertilizers may cause some temporary changes in the biological processes essential for soil fertility, but the changes are neither permanent nor irreversible. The benefits of fertilizer use, associated with the increased availability of an element essential for plant growth, far outweigh the temporary inhibition of certain soil microorganisms.

RECOMMENDATIONS

1. Agricultural extension programs should be designed to educate farmers to apply fertilizers with full consideration of the nitrogen release characteristics of the soil, climate, the expected timing of plant uptake demands, and anticipated temperature and rainfall in the watershed. Since information is lacking in some of these areas, additional research should be sponsored by appropriate federal and state agencies.

2. Means for maximizing plant uptake of nitrogen should be developed.

3. Tests for plant-available soil nitrogen that can be used in all major agricultural areas of the United States should be developed.
4. The cost to society of the loss of most of another increment of fertilizer nitrogen to waterways should be determined. This would be in contrast with the traditional determination of the economic benefit of using an additional increment of fertilizer for crop production.
5. The efficacy of foliar applications of nitrogen fertilizer in crop production should be improved.
6. The long-term benefits of incorporating organic manures into soils as a means of recycling nitrogen should be assessed.
7. Means of minimizing use of fertilizer while maximizing crop yields and quality in irrigation farming should be found.
8. Crop varieties that will scavenge the soil for inorganic nitrogen and convert it to protein without accumulating nitrate should be developed.

Hazards of Nitrate, Nitrite, and Nitrosamines to Man and Livestock

Changing patterns in agricultural practice, food processing, urbanization, and industrialization have had an impact on the accumulation of nitrate in the environment. There is an ever increasing use of nitrogen fertilizers in crop production, particularly with corn, vegetables, other row crops, and forages. Livestock feeding and poultry operations are expanding, causing higher concentrations of nitrogenous wastes in small areas of land. There is more open land with no cover crops, and herbicides are providing weed-free fields. Nitrate and nitrite are used extensively for color enhancement and preservation of processed meats and meat products. Some of these developments are associated with the increased exposure of man and animals to high nitrate levels in food, feed, and water. At the same time, increasingly accurate diagnostic techniques make greater precision possible in the assessment and inventory of disorders arising from the exposure.

SOURCES OF EXPOSURE

NITRATE AND NITRITE IN HUMAN FOOD

Vegetables Nitrate occuring naturally in the human food supply may enter the human diet through food sources or water. This is especially true of vegetables, such as beets, spinach, broccoli, celery, lettuce, radishes, kale, mustard greens, and collards—all of which may accumulate large quantities of nitrate. Representative sources are given in Table 10. Nitrite, on the other hand, is generally found in very small quantities in the fresh and processed commodity (Table 11).

Until recently, there were no recorded instances of methemoglobinemia resulting from the consumption of vegetables (Brown and Smith, 1966; Jackson *et al.*, 1967). However, several cases of methemoglobinemia and one death have been reported recently from Europe (Committee on Nutrition, 1970; Fassett, 1966). In these instances, the illness resulted from the use of home preparations of pureed samples of fresh spinach held under questionable storage conditions and later fed to infants 2 to 10 months of age. Commercially prepared baby foods have been implicated in only one case of methemoglobinemia in the United States. In this instance, which occurred in 1970, a 1-month-old infant became ill

TABLE 10 Nitrate Content of Vegetables Grown in 1963 and 1964 and Vegetables Purchased in Columbia, Missouri, Stores in 1964[a]

	NO_3-N Content (% dry weight)		
Vegetable	Field Grown[b] (1963)	Field Grown[b] (1964)	Purchased[c] (1964 range)
Radishes (red)	0.53–1.2	0.8 –1.9	0.39–1.50
Beets (red)	–	0.19–0.78	0.09–0.84[d]
Turnips, tops	0.25–0.85	–	0.03–0.76
Carrots	0.02–0.05	0.02–0.05	0.0 –0.13
Lettuce, leaf	0.08–0.5	0.09–0.60	0.02–1.06
Spinach	–	0.09–0.24	0.07–0.66
Kale	0.30–1.02	–	–
Mustard	0.46–0.98	–	–
Sweet corn	–	–	0.01
Cabbage	–	–	0.01–0.09
Broccoli	–	–	0.01–0.09
Cauliflower	–	–	0.0 –0.31
Celery	–	–	0.11–1.12
Green beans	–	–	0.04–0.25
Squash	–	–	0.09–0.43
Cucumbers	–	–	0.0 –0.16
Tomatoes	–	–	0.0 –0.11

[a]From Brown and Smith (1967).
[b]Low values were from plants grown on soil receiving no nitrogen fertilizer, and high values were from plants grown in soil receiving 400 lb N/acre (450 kg/ha).
[c]Includes both locally grown and shipped-in supplies.
[d]Baby foods.

TABLE 11 Prepared Infant Foods Containing More Than 20 ppm of Nitrate Nitrogen[a]

Food	No. of Samples	NO_2-N (ppm)		NO_3-N (ppm)	
		Range	Avg.	Range	Avg.
Mixed vegetables	2	0–2	0.8	21–24	22
Carrots	8	0–2	0.7	15–38	23
Green beans	3	0–1	0.8	16–71	37
Garden vegetables	5	0–2	0.7	19–62	41
Squash	5	0–1	0.7	10–93	64
Wax beans	2	1–2	2	73–129	101
Beets	6	0–2	0.8	144–492	222
Spinach	5	0.3–1	0.5	244–379	312

[a] Source: Kamm *et al.* (1965).

after being fed commercially packed strained beets; the infant recovered and had no further difficulty. The extensive use of these foods, especially spinach and beets (which are often particularly rich in nitrate), without deleterious effects provides circumstantial evidence of their harmlessness. Over 350 million jars of canned spinach were used in the United States and Canada during a 20-yr interval, with no proven instance of methemoglobinemia (Committee on Nutrition, 1970). If these products are injurious, one would expect that several instances of methemoglobinemia would have been reported. The one recent case is noteworthy both for its occurrence and its rarity. In addition, vegetables rich in nitrate are less than 3 percent of the volume of processed baby foods, and this percentage is decreasing (G. A. Purvis, personal communication).

Extensive studies have established the variables affecting nitrate and nitrite content of vegetable crops. It is known, for example, that the nitrate concentration is a function of the amount of fertilizer applied, and the source of nitrogen is not significant in determining nitrate levels when it is applied as a side-dressing. It is also evident that conditions that slow the rate of growth—such as drought, deficiencies of nutrients other than nitrogen, low light intensity, or insect damage—may result in higher nitrate levels. Furthermore, not only is there a great variation in the amount of nitrate that plant species accumulate but there is also a wide range in the nitrate levels in samples of the same commodity (Barker *et al.*, 1971; Brown and Smith, 1966; Jackson *et al.*, 1967; Smith, 1970; Walters, 1970). Of interest, too, in view of the changes in fertilizer application rates in recent years, is the fact that the nitrate content of present-day vegetables is not appreciably different from that of vegetables produced half a century ago (Jackson *et al.*, 1967).

People in the United States apparently run little risk of either acute or chronic injury as a result of eating fresh or processed vegetables. It has been pointed out that no cases of poisoning had been noted in this country or in Canada (Phillips, 1971). Reports from Europe suggest that a potential for nitrate toxicity still exists (Committee on Nutrition, 1970; Fassett, 1966), and the single U.S. case in 1970 from commercial baby food cannot be ignored. Hence, attention should be given to factors controlling nitrate levels in the production and handling of vegetables that accumulate nitrate and are used extensively. For example, it is known that nitrate accumulation in spinach can be reduced by applying nitrogen fertilizer as required by the crop, rather than as a single heavy application (Barker *et al.*, 1971). Excessive nitrogen fertilization should be avoided. Moreover, prudence would dictate that nitrate-rich vegetables should not be fed to infants during the first few months of life.

Nitrite is not commonly found in packaged fresh spinach, but it may appear after storage at room temperatures or in the refrigerator. Similarly, nitrite, although not found in canned spinach when opened, may form in open cans of spinach or in frozen spinach when allowed to thaw (R. L. Minotti, unpublished data). Therefore, prolonged storage of nitrate-rich products under conditions favorable for nitrite accumulation should be avoided.

Meat Nitrate has been used since ancient times to preserve and cure meat. In addition, it imparts a red color and thus serves an aesthetic purpose. Because the chief contributor to the red color is nitrite, normal practice is to use a mixture of nitrite and nitrate. Nitrite also inhibits microorganisms that might grow in the food product. Ham, corned beef, sausages (including wieners), and some preserved fish are treated with nitrite and nitrate. The Food and Drug Administration has established tolerance limits of 200 ppm of nitrite and 500 ppm of nitrate as the final concentrations that can be added to cured meat and some preserved fish. These values are as nitrite and nitrate, not as nitrogen. Although occasional poisonings have been associated with treated foods, they have arisen through accidental addition of excessive amounts of nitrate–nitrite mixtures to the foods.

The volume of meat products so treated under federal inspection illustrates the widespread acceptance of cured meat products by consumers. In 1969, 230 million lb of beef and 3.6 billion lb of sausage (including headcheese and other food products) were processed in this country (American Meat Institute Foundation, 1970). The use of nitrite plus nitrate for curing meat has plateaued, and no appreciable increases have been noted in recent years.

The nitrate content of selected cured meats ranges from 0 to 600 ppm and the nitrite content from 2 to over 200 ppm. The amount of nitrite present in a product is generally reduced during heat processing, and there is apparently a continuous loss of nitrite during storage. A further loss occurs if the products are heated prior to consumption.

NITRATE AND NITRITE IN LIVESTOCK FEED

Most crop plants grow more rapidly when supplemented with nitrogen in the form of fertilizer; because different species and strains vary in their capacity to change the nitrate that they assimilate to protein, nitrate may accumulate in varying amounts. Under normal cropping practices and environmental conditions, plants contain relatively low levels of nitrate. If little or no nitrate is added to soil, the amount in plants usually ranges from 0.0050 to 0.05 percent of nitrate nitrogen (Table 12). The nitrate-accumulating plants (many of which are

TABLE 12 Nitrate Concentrations in Plants Grown on Soil with Little or No Added Nitrate[a]

Plant	Country	Nitrate Nitrogen (%)
Pasture hay	Netherlands	0.0045
Clover hay	Netherlands	0.0068
Greenstuff silage	Netherlands	0.0045
Grass (fresh)	Netherlands	0.05
Grass (dried)	Netherlands	0.065
Grass (silage)	Netherlands	0.026
Sugar beet leaves	Germany	0.047
Turnip	Germany	0.068
Turnip leaves	Germany	0.072
Maize	Germany	0.063
Cocksfoot	Germany	0.044
Grass (sewage irrigated)	Germany	0.039
Grass	United Kingdom	0.01
Rape	United Kingdom	0.05–0.12
Rape	United Kingdom	0.03
Green oats (stems)	United States	0.011

[a]Source: Becker (1967).

weeds) include barley, wheat, corn, sugar beet, sunflower, turnip, pigweed, thistle, lamb's-quarter, bindweed, nightshade, ragweed, certain algae, and sometimes alfalfa (Muenscher, 1961). Nitrate tends to accumulate in vegetative portions of the plant; little appears in grain.

Some of the factors that lead to the accumulation of nitrate in plants are dry hot seasons, heavy manure treatments, and insufficient levels of phosphorus or other plant nutrients required for normal plant metabolism. Others are sudden changes in temperature, frost, shading of plants, insect infestations, lack of balance among nutrients in soil, and certain herbicides. The stage of maturity of the plant also affects its nitrate content.

The amount of nitrate in plants increases when too much nitrogen is supplied. For example, 0.17–0.37 percent of nitrate nitrogen has been found in Sudan grasses and in sorghum–Sudan grass crosses grown with no nitrogen fertilization, whereas the levels ranged from 0.36 to 0.50 percent when 112 kg nitrogen/ha was applied (G. E. Smith, personal communication). On the other hand, corn, sorghum, and oats were reported to contain 1.95–2.50 percent of nitrate nitrogen before chemical fertilizers were in common use (Bradley *et al.*, 1940; Mayo, 1895; Pease, 1896). In addition, large amounts of nitrogen can be applied without causing nitrate accumulation (Alexander *et al.*, 1961; Stallcup *et al.*, 1960).

Nitrate was reported in the 1950's to be present in soybean meal, but the source was accidental fertilizer contamination, not the plant. There are still occasional reports that concentrate portions of the ration contain nitrate, but the source of the nitrate is not known (Marrett and Sunde, 1968). Molasses from cane and sugar beets, especially cane grown on rich soils or highly fertilized sugar beets, can be high in nitrate. By-products from sugar processing may also contain nitrate.

Nitrite is not a common component of growing plants. When a plant is bruised, however, nitrite may accumulate as a result of the release of plant enzymes or proliferation of bacteria (Olson and Moxon, 1942), and as much as 13 percent of the nitrate may be converted to nitrite in a day (Barnett, 1953). This reduction of nitrate to nitrite is probably favored by high moisture content and high temperature. Moreover, any nitrate-containing food or feed processed by bacterial fermentation may, at some stage, contain appreciable nitrite. Thus, silage or "mash" prepared with water high in nitrate commonly contains some nitrite, especially in the early stages of fermentation.

When fed with proper management, nitrate may serve as a nonprotein nitrogen source for ruminants. Several studies have revealed that lambs tolerate up to 4 percent of potassium nitrate in low-protein hay rations supplemented with limited quantities of concentrates. The added nitrate increases nitrogen digestibility, increases levels of nitrate in the blood, and improves nitrogen balance. The value of nitrate as a nonprotein nitrogen source has been demonstrated by Lewis (1951b), Sapiro *et al.* (1949), and numerous other investigators. There is also considerable interest in nitrate and other nitrogen sources, such as urea, as feed supplements for ruminant animals.

NITRATE ACCUMULATION IN WATER

Evidence of increases in the concentration of nitrate in some surface waters has been presented (see "Changes in Storage of Nitrogen," p. 11). Two aspects of this concentration are causes of concern: Levels in excess of 0.3 mg of nitrogen per liter of water are believed to contribute to excessive algal growth; concentrations of nitrate nitrogen in excess of 10 mg per liter of water exceed Public Health Service recommended limits for potable water.

In Illinois, some surface waters do contain nitrate in excess of 10 mg per liter of nitrate nitrogen, and most contain more than 0.3 mg per liter. In 1962, the U.S. Geological Survey examined water supplies of the 100 largest cities in the United States and found a median concentration of 0.16 mg per liter of nitrate nitrogen and a maximum of 5.5 mg per liter (U.S. Geological Survey, 1962). More recently, the U.S. Public Health Service examined 969 water supply systems serving about 18 million people in eight Standard Metropolitan Statistical Areas and the state of Vermont (U.S. Public Health Service, 1970). Of 2,595 samples analyzed, 53 exceeded the limit of 10 mg per liter of nitrate nitrogen; the maximum observed was 29 mg per liter. Nineteen of the 969 systems exceeded 10 mg per liter of nitrate nitrogen. According to this report, most of the high nitrate levels occurred in groundwater in California. This most recent analysis indicates that most supplies of potable surface water conform with U.S. Public Health Service drinking-water standards, although many surface waters contain more nitrogen than is believed necessary for algal growth.

Although there are few records on trends in nitrate concentrations in groundwater, many groundwater supplies may contain nitrate in excess of 10 mg of nitrate nitrogen per liter. Most reports of particularly high nitrate levels in groundwater supplies are from rural areas, and agricultural sources of the nitrogen are frequently suspected. However, there is little evidence of any causal relationship; rather, it appears that groundwater is more likely to become contaminated than surface water, and in rural areas more drinking water comes from wells than from surface waters. For this reason, nitrate concentrations in groundwater will be considered here rather than a comparison of rural with urban supplies.

A summary of analyses of 8,844 wells in Illinois is shown in Table 13. Of the samples from wells 0–8 m in depth, 28 percent exceeded the Public Health Service recommended limit of 10 mg per liter and 13 percent contained more than 20 mg per liter (Larson and Henley, 1966). The results also clearly show a decrease in nitrate content as well depth increases. A summary of analyses of over 5,000 wells in Missouri showed that 27 percent of the wells exceeded 10 mg per liter, but no relationship to soil type was evident (G. E. Smith, personal communication). The Illinois State Water Survey found that 81 percent of 221 dug wells and 34 percent of 38 drilled wells in Washington County, Illinois, contained more than 10 mg per liter of nitrate nitrogen. By contrast, most of 72 farm ponds had less than 1.1 mg per liter and none had more than 2.9 mg per liter, indicating that groundwaters are generally richer in nitrate than pond waters. In addition, of 250 wells in Wisconsin examined twice monthly for over a year, 71 percent exceeded 10 mg per liter of nitrate nitrogen at least once and about 45 percent exceeded 10 mg per liter throughout the sampling period (Crabtree, 1970).

Particularly striking are the observations in Nassau County, New York, where over 370 wells supply about 1½ million people. By

TABLE 13 Relationship between Depth of Wells and the Nitrate Content of the Water[a]

Depth of Wells (m)	Number of Analyses	Concentration (mg/liter as N)[b]			
		0.2	2	10	20
0–8	480	87	56	28	13
9–15	926	80	40	20	10
16–30	1,568	64	18	5	1.8
31–60	2,042	61	11	3	0.7
over 60	3,828	55	5	0.6	0.1

[a]Source: Illinois State Water Survey.
[b]Percentages of analyses having nitrate equal to or greater than each of the four concentrations shown.

1969, the nitrate nitrogen content of the water in 20 of these wells was in excess of 10 mg per liter (Smith and Baier, 1969), and they were either shut down or blended with low-nitrate water. Moreover, certain water supplies in southern California have exceeded 10 mg/liter of nitrate nitrogen since 1935.

With the exception of the observations in Illinois showing decrease of nitrate concentrations with increasing well depth, none of these studies has revealed clearly defined relationships between nitrate content and type of well, soil type, geological formation, or land use. However, it is evident that concentrations of nitrate in groundwater supplies in many parts of the country exceed the Public Health Service recommended limit of 10 mg per liter of nitrate nitrogen.

NITROSAMINES IN FOOD

Nitrosamines have recently received considerable attention because of their possible carcinogenic, teratogenic, and mutagenic properties. Information on their occurrence in food is being accumulated, but it remains scanty and inconclusive. Although nitrosamines are found in some nonprocessed food products, they occur more frequently in processed foods. The precise levels of nitrosamines present in certain foods is difficult to determine because of the lack of a sensitive and reliable analytical technique for detecting them.

Nitrate, nitrite, and secondary and tertiary amines are precursors of nitrosamines and, as stated on pages 46 and 47, nitrate and nitrite occur in processed meats, produce, and canned goods. Some investigators believe that processed meat products provide an unfavorable climate for the formation of nitrosamines, because myoglobin or hemoglobin neutralizes the effect of the highly reactive nitrosating agent, NO^+, produced by the nitrate and nitrite additives. This neutralization is brought about by a reaction of the nitrosyl cation with the ferrous-iron of the myoglobin or hemoglobin. Since these neutralizing conditions are presumably not present in fish or cheese, the possibility of forming nitrosamines from nitrate and nitrite seems greater in these two food sources. Fish also have natural supplies of different amines, and nitrate and nitrite added to fish flesh can react to produce nitrosamines (Ender *et al.*, 1964). Furthermore, nitrosamine precursors may be converted to nitrosamines in the stomachs of mammals (Sander, 1967; Sander *et al.*, 1968; Sen *et al.*, 1969a).

The data reporting the actual testing of these food products are scanty. However, using thin-layer chromatography to assay for nitrosamines, Ender and Ceh (1967, 1968) found up to 40 μg/kg in various smoked fish, up to 6 μg/kg in various smoked meats, and up to 30 μg/kg in some species of mushrooms. A few tests performed on other food products, including flour, have given conflicting data (Ender and Ceh, 1967; Mohler and Mayrhofer, 1968; Sen *et al.*, 1969b; Marquardt and Hedler, 1966); for example, the high levels reported in flour by Marquardt and Hedler (1966) were not found by Thewlis (1967). On the basis of these studies, it is evident that more thorough tests will be required to resolve the question of the occurrence and precise levels of nitrosamines in food products.

Recent unpublished studies by the U.S. Food and Drug Administration and the U.S. Department of Agriculture show that traces of dimethylnitrosamine are found, although very infrequently, in frankfurters, a dried beef product, and cured ham. Nitrosopyrrolidine was also noted in bacon after, but not before, cooking. The presence of these substances may be due to the addition of nitrite in concentrations above the level permitted by the Food and Drug Administration.

The research performed to date clearly shows that present analytical methods—colorimetry, thin-layer chromatography, gas chromatography, and mass spectrometry—are inadequate for assessing the occurrence and quantities of nitrosamines in foods. As a result, although nitrosamine values have been determined for certain food products, their precise levels are questionable. Therefore, a critical need exists for the further development of methodology, so that accurate and reliable measurements can be made.

METHEMOGLOBINEMIA FROM NITRATE AND NITRITE

INFANTS AND ADULTS

Methemoglobinemia, which can arise from nitrite or indirectly from nitrate, can lead to difficulty through impairment of oxygen transport in the blood. It has been recognized as a health problem for many years. A review by Lee (1970a) gives an extensive bibliography of methemoglobinemia associated with nitrate and nitrite.

Poisoning in infants from nitrate in well water was first reported in 1944, and it is said that some 2,000 cases have now been reported (Shuval *et al.*, 1970). Table 14 gives data on about 350 cases in the United States and about 1,000 in Europe in which the cause was reported to be nitrate or nitrite in well water or foodstuffs. Fatalities resulting from these cases totaled 41 in the United States and about

TABLE 14 Reported Cases of Methemoglobinemia in the United States and Europe Resulting from Nitrate or Nitrite in Well Water or Foodstuffs

No. of Cases	No. of Fatalities	Reported Cause	Years	Reference
United States				
278[a]	39	Well-water nitrate	1945–1950	Walton (1951)
40	0	Well-water nitrate	1952–1966	Bailey (1966)
10[a]	0	Well-water nitrate	1960–1969	13 state health departments[b]
12	1	Nitrite added to meat[c]	1955	Orgeron *et al.* (1957)
3	1	Nitrite added to fish[c]	1959	Singley (1962)
Europe				
1,000[a]	80[a]	Well-water nitrate	1948–1964	Knotek and Schmidt (1964); Simon *et al.* (1964); Downs (1950); Aussannaire *et al.* (1968); Ewing and Mayon-White (1951)
15	1	Nitrite and nitrate in spinach	1959–1965	Sinios and Wodsak (1965)

[a] Approximate.
[b] Personal communications.
[c] Accidental overaddition.

80 in Europe. Most of the cases were in the period 1945–1950. In the United States, only one case of an infant developing methemoglobinemia as a result of drinking water obtained from a public water supply has been reported (Vigil *et al.*, 1965).

Nitrite that reaches the bloodstream reacts directly with hemoglobin to produce methemoglobin, with consequent impairment of oxygen transport. Nitrate itself is relatively nontoxic to mammals, being readily absorbed and readily excreted. Under certain circumstances, however, nitrate can be reduced in the gastrointestinal tract to nitrite. Thus, the methemoglobin response is dependent upon either preformed nitrite or nitrite formation in the body.

Apparently the reaction of nitrite with hemoglobin is inconsequential in adults, but it can be troublesome in infants, particularly in infants under 3 months of age. This striking difference in response is based on the following circumstances: (a) Fetal hemoglobin, which is still present in the newborn, may be more readily oxidized to methemoglobin than adult hemoglobin; (b) infants are deficient in two enzymes in their red blood cells, methemoglobin reductase and diaphorase, which convert methemoglobin to hemoglobin; (c) illness involving the gastrointestinal system in infants may permit the bacteria responsible for the reduction of nitrate to nitrite to move higher in the gastrointestinal tract, increasing the likelihood of nitrite formation; and (d) the stomach pH in infants is less acidic than in adults, encouraging bacterial growth in the stomach and upper intestine (Miale, 1967; Walton, 1951). All the reported cases of methemoglobinemia associated with well water have been in infants.

In addition to arising from nitrate in water, methemoglobinemia may be brought about by consuming vegetables and baby food. Poisoning has also been reported in the United States from the accidental over-addition of nitrite during the curing of meat and fish (see Table 14).

Methemoglobin is normally present at levels of 1–2 percent of the total hemoglobin in blood. Only at levels of about 10 percent are clinical symptoms normally detectable. Concentrations of 30–40 percent are compatible with life but will normally lead to anoxic symptoms. Death occurs at levels of 50–75 percent.

The relationship between maternal and fetal methemoglobin levels is unknown. However, methemoglobin levels seem to be significantly higher in premature infants than in newborn mature infants, older children, or adults (Kravitz *et al.*, 1956). Whether high values are present in the prenatal state is unknown.

There is some evidence that the presence of methemoglobin in the blood alters the affinity of oxygen for hemoglobin in a manner analogous to that of carbon monoxide, tending to make the release of oxygen from the blood to the tissues somewhat more difficult. However, this effect is considerably less prominent with methemoglobin than it is with carboxyhemoglobin (Bodansky, 1951).

The disease in its uncomplicated acute form is usually readily diagnosable: It occurs promptly after the nitrate or nitrite is ingested; it generally makes itself evident through the characteristically cyanotic appearance of the infant ("blue baby disease"); and it is easy to measure the methemoglobin in the blood at these levels. Treatment, as with methylene blue or ascorbic acid (Miale, 1967), is readily available and quite generally understood. The possibility must be recognized, however, that methemoglobinemia in infants may be strongly dependent on the presence of gastrointestinal upsets and, short of major cyanosis, might go unnoticed in an infant already ill.

Methemoglobinemia in which there is less than about 10 percent of methemoglobin—the so-called subclinical range—has been generally regarded as of no medical importance. However, a report by Petukhov and Ivanov (1970) describes the slowing of conditioned motor reflexes in response to auditory and visual stimuli in 39 children (age 12–14) whose drinking water contained 26 mg/liter of nitrate nitrogen. Their reflexes were compared with those of a control group whose water contained 2 mg/liter of nitrate nitrogen. The increase in reaction time was statistically significant with both kinds of stimuli. The average methemoglobin in the blood of a randomly chosen sample of the group on the high-nitrate water averaged 5.3 percent (10 children), compared with 0.75 percent for a randomly selected sample of the group on the water with less nitrate (11 children). This work has not been confirmed, and the issue raised obviously needs further examination.

It has been reported that low-level exposure to carbon monoxide produces detectable but reversible effects on the ability to judge time intervals (Beard and Wertheim, 1967). These observations have some parallelism with the report noted above describing slowing of conditioned motor reflexes in children with methemoglobinemia. These behavioral effects reported for methemoglobinemia and carbon monoxide are subtle and require special techniques for their detection. Presumably they are temporary functional alterations and are reversible.

In all instances, infant poisoning associated with water has arisen from nitrate in the water used for preparing formulas. In some instances, the nitrate was presumably further concentrated by boiling.

Only a crude correlation exists between the concentration of nitrate in water and the frequency of clinical cases of infantile methemoglobinemia. The preponderance of cases have occurred where water sources to which the poisonings were attributed exceeded 22 mg/liter of nitrate nitrogen. In two studies, however, it was reported that 3.0 and 4.4 percent, respectively, occurred with nitrate nitrogen concentrations below 9 and 11 mg/liter (Table 15), equivalent to 40 and 50 mg NO_3/liter.

Several studies conducted specifically on subjects without symptoms show a rough correlation between nitrate content of the water supply and infant methemoglobinemia. As can be seen in Table 16, all of the methemoglobin levels are quite low in the individuals without symptoms and barely above control levels. The lack of closer correlation between nitrate in the water and methemoglobin levels is not surprising considering the confusing factors that could intrude, particularly delays between diagnosis and blood and water analyses and uncertainty as to the identification of the source of water. Diskalenko (1968) also reports a correlation of nitrate in water with methemoglobinemia.

TABLE 15 Distribution of Reported Cases of Infantile Methemoglobinemia by Nitrate Concentration in Water

	From Simon *et al.* (1964)		From Sattelmacher (1962)	
	No.	%	No.	%
Reported cases	745	100	1,060	100
Deaths	64	8.6	83	7.8
Nitrate concentration in water (mg/liter)[b]				
Unknown	496	66.5	593	56.0
0–40	–	–	14	3.0[a]
0–50	11	4.4[a]	–	–
41–80	–	–	16	3.4[a]
50–100	29	11.8[a]	–	–
81–100	–	–	19	4.1[a]
>100	209	83.8[a]	418	89.5[a]

[a] Percentage of cases with nitrate concentration known.
[b] Values are as nitrate (NO_3) and not nitrogen.

The nitrite concentration in water is normally very low, rarely rising to levels of 1 mg of nitrite nitrogen per liter, but nitrate is often very much higher. The Public Health Service standard of 10 mg/liter of nitrate plus nitrite nitrogen (equivalent to 45 mg/liter of nitrate) has been adopted in many countries. In a few instances, this level of nitrate has been exceeded in public water supplies in the United States, and it is frequently exceeded in shallow wells. Most of the infant poisonings have occurred in rural areas where well water was used.

TABLE 16 Relation of Nitrate Concentration in Water Used in Diet and Methemoglobinemia in Infants without Known Disease

Nitrate Nitrogen in Water (mg/liter)	Number of Infants	Methemoglobin (%)	References
0–11	2,038[a]	0.95–1.0	Shuval *et al.* (1970)
11–15	–	1.3 –1.4	
0–11	80	1.0[b]	Simon *et al.* (1964)
		0.8[c]	
11–22	38	1.3[b]	
		0.8[c]	
>22	25	2.9[b]	
		0.7[c]	

[a] Total number in the study.
[b] 0–3 months.
[c] 3–6 months.

As noted above ("Nitrate and Nitrite in Human Food," p. 46), nitrate and nitrite are extensively used in processed meats and in some fish, and the accidental addition of excessive amounts of nitrite has led to poisoning. The tendency of some vegetables, notably spinach, to accumulate nitrate and occasionally to contain nitrite during storage is discussed in the same section.

An attempt has been made to develop means for predicting the extent of methemoglobinemia from nitrate intake over a range of ages (Table 17). Winton *et al.* (1971) took infants 1 to 3 months old as the most sensitive group. For a 3-month-old infant weighing 5.4 kg, they estimate that 2 mg of nitrate nitrogen/kg of body weight, or about 11 mg, would be required to produce 10 percent methemoglobinemia (the upper limit of the "subclinical range"). This level would be reached by the intake of a little less than 1 liter of water contaminated to the extent of 10 mg of nitrate nitrogen or 45 mg of nitrate/liter, the Public Health Service standard. These estimates in the infant assume a high (80 percent) efficiency in the bacterial reduction of nitrate, a mole-for-mole ratio in the reaction of nitrite with hemoglobin, and delivery of the daily dose in six fractions every 4 hr.

It is possible that treated meats, such as wieners, might be fed to infants. On the basis of the susceptibility of a 3-month-old infant as shown in Table 17, it would appear that about 20 g per day of meat containing the allowable amounts of nitrate and nitrite could produce a 10 percent methemoglobinemia. Winton *et al.* (1971) confine their estimates for the adult to nitrite because adults are relatively insensitive to nitrate. It would appear that a daily intake by the adult of about 1 mg of nitrite (0.3 mg of nitrite nitrogen)/kg of body weight, or 70 mg for a 70-kg adult, is needed to produce the same endpoint of 10 percent methemoglobinemia. This amount would be contained in 350 g of ham or other treated meat containing the maximum allowed by the Food and Drug Administration.

LIVESTOCK

Animals have become ill or died after consuming feed or water containing high concentrations of nitrate or nitrite. Natural outbreaks may result in the sudden death of 10–30 percent of the animals in a herd, and losses in drought-affected areas and from early spring grazing may number in the thousands in a single year. While the number of animals lost is a small percentage of the livestock in a state, the loss of 10 percent of a herd is a disaster for the owner. Fortunately,

TABLE 17 Hypothetical Dose of Nitrate Capable of Converting 10 Percent Hemoglobin to Methemoglobin, by Age[a]

Age	Total Body Wt (kg)	Hemoglobin (g/100 ml)	Blood Volume (ml)	Hemoglobin/Body Wt (g/kg)	Nitrite[b] (mg/kg)	Nitrate[b] (mg/kg)	Nitrate per 24 hr[b] (mg/kg)
1 mo	3.6	16	500	22.2	1.6	2.7	16.2
2 mo	4.5	12	550	14.7	1.0	1.7	10.2
3 mo	5.4	11	600	12.2	0.9	1.5	9.0
5 mo	6.8	11	750	12.2	0.9	–	–
10 mo	9.5	12	1,000	12.7	0.9	–	–
3 yr	14.1	13	1,500	13.8	1.0	–	–
5 yr	20.4	13	2,250	14.4	1.0	–	–
Adult	70.0	14	6,300	12.6	0.9	–	–

[a] Source: Winton *et al.* (1971)
[b] For 10 percent conversion of hemoglobin to methemoglobin.

methods have been developed for predicting outbreaks, for detecting the buildup of nitrate, and for altering management practices to cope with detected increases in nitrate concentration in feed and water supplies. The availability of information makes it unlikely that acute outbreaks of the magnitude observed in the Midwest in the mid-1950's will occur again.

The chronic effects of nitrate and nitrite are not as well documented as are the acute effects. Several patterns of reduced productivity have been identified as being the result of long-term exposure to nitrate or nitrite. Healthy animals can adapt to nitrate. Eventually they can tolerate small amounts, and this occurs without a detectable buildup of toxic substances.

There is a lack of agreement on what constitutes a toxic dose for livestock. Part of the confusion results from the fact that investigators have expressed their findings in different ways; moreover, the nitrate or nitrite is applied by different procedures, and animals of different ages and nutritional states have been used in different environments. In Tables 18, 19, and 20, the data have been placed on similar bases by expressing toxicity in terms of the amount of nitrate or nitrite nitrogen per unit of body weight. Since food intake has not always been reported, it has sometimes been impossible to calculate the total dose.

The process leading to methemoglobinemia is initiated by the reduction of the nitrate consumed by the animal to nitrite. Nitrite is an intermediate in the conversion of nitrate to ammonium by both bacteria and plant tissues. In certain instances, nitrate does not appear in detectable amounts during the reduction. The nitrite is readily absorbed, and the ruminants are quite sensitive to nitrate in feed and water because of the nitrite buildup. Monogastric animals are sensitive to administered nitrite. The newborn pig is probably as sensitive as the human infant.

If the animal consumes large quantities of nitrite or if the further reduction of the nitrite formed from nitrate does not occur, nitrite will be absorbed from the rumen or the midportions of the intestine, and methemoglobin will be formed. As much as 70 percent of the hemoglobin can be converted to methemoglobin in 2–5 min. It appears that usually about 75 percent of the hemoglobin can be converted to methemoglobin (Sokolovski and Pavlova, 1961). Blood containing methemoglobin usually has a chocolate-brown color, although blood with a brilliant red color has also been reported (Case, 1957). Other compounds, including nitric oxide hemoglobin and nitroso hemoglobin, may be formed.

The toxicity resulting from a given quantity of nitrate or nitrite is determined by the specific reduction products that become available for absorption and the activity of methemoglobin reductase. Rations containing good forage support a strongly reducing environment and are more protective against nitrite toxicity than rations of poor quality (Holtenius, 1957; Pfander *et al.*, 1957). Toxicity will not appear if the nitrate is reduced to ammonium without nitrite accumulating; ammonium is used by bacteria in the gastrointestinal tract for growth.

TABLE 18 The Toxicity of Nitrate Administered by Feed, Drench, or Intravenous Injection

Animal Tested	Nitrate Dose (mg N/kg body wt)	Effect	Reference
Sheep	83	Splenomegaly	Miyazaki (1967)
	99 + hay	1 death	Buchman *et al.* (1968)
	112	>5 g methemoglobin/ 100 ml blood	Emerick *et al.* (1965)
	112	30% weight reduction	Miyazaki (1967)
	[illegible]	Minimum lethal	Clarke and Clarke (1967)
	150	Death, perhaps abortion	Davison *et al.* (1965)
	0.67% of diet[a]	Depressed gain	Sokolowski (1966)
	14 mg/100 ml blood	6 g methemoglobin/ 100 ml blood	Emerick *et al.* (1965)
	175	LD_{50}[b]	W. H. Pfander (unpublished data)
	350[c]	Diuresis	Pfander *et al.* (1957)
Dairy cows	45	Death in 3–13 days	Simon *et al.* (1959)
	56	None	Jones *et al.* (1966)
Cattle	62	50% of hemoglobin changed to methemoglobin	Crawford *et al.* (1966)
	63	Death on day 2	Simon *et al.* (1959)
	74	LD_{50}[b]	Bradley *et al.* (1940)
	226	LD_{50}[b]	Crawford *et al.* (1966)
Heifers	86	Death	Winter and Hokanson (1964)
	100	One abortion	Davison *et al.* (1964)
	150	Two abortions, two deaths, low conception	Davison *et al.* (1964)
	150	No effect on reproduction, cyanosis	Winter and Hokanson (1964)
Dog	570[c]	Diuresis, chloride loss	Greene and Hiatt (1954)
Rat	800[c]	LD_{50}[b]	Wright and Davison (1964)

[a] Intake not reported.
[b] LD_{50} is that dose which will cause death in 50 percent of the animals administered the dose. This is a useful pharmacological concept, but it is of no direct value to the livestock producer.
[c] Administered by injection.

TABLE 19 The Toxicity of Nitrite Administered in Feed, Intrarumenally (IR), or by Intravenous (IV) or Intraperitoneal (IP) Injection

Animal Tested	Nitrite Dose (mg N/kg body wt)	Administered	Effect	Reference
Sheep	15	IV	Death in 3 hr	Pfander *et al.* (1957)
	30	Fed	Minimum lethal dose	Clarke and Clarke (1967)
	34	Unknown	Calculated lethal dose	Burden (1961)
	66	IR	Death in 2–3 hr	Sinclair and Jones (1967)
	50	Unknown	Acute lethal dose	Case (1970)
	69	In 2 feeds	None	Sinclair and Jones (1967)
	78	IP	Death in 2 hr	Diven *et al.* (1964)
	80	IR	Death	Lewis (1951a); Holst *et al.* (1961)
Cattle	20 mg	IR	LD_{50}	Stormorken (1953)
	45 mg	Fed at intervals	LD_{50}	Stormorken (1953)
	50	[a]	Acute lethal dose	Case (1970)
Horses	50	[a]	Acute lethal dose	Case (1970)
Pigs	12	[a]	Minimum lethal dose	Clarke and Clarke (1967)
	18	[a]	Calculated lethal dose	Burden (1961)
	21	[a]	LD_{50}	Nelson (1966)
	23	[a]	Death	Winks *et al.* (1950)
	24	[a]	Acute lethal dose	Case (1970)
Rabbit	0.6	[a]	Hypotension, tachycardia	Iacovoni *et al.* (1968)
	18	[a]	Calculated lethal dose	Burden (1961)
Dog	64	[a]	Calculated lethal dose	Burden (1961)

[a] It is assumed that all the nitrite reached the blood.

Animal blood contains active methemoglobin reductase, an enzyme whose activity varies with species, individual, and age. Sheep reduce methemoglobin rapidly, cattle more slowly, and pigs and horses still more slowly (Bartik, 1964; Smith and Beutler, 1966). Young adults have a greater reductive capacity than old animals, and very young animals may have a limited capacity.

Gibson (1943) showed that ascorbic acid is useful for reducing methemoglobin. Methylene blue, however, is usually the treatment of choice, and it will lower the methemoglobin content from 70 percent to as low as 5 percent in a relatively short time.

In field cases, livestock losses are greatest about 3 to 5 days after the animals are first exposed to feed containing high levels of nitrate. However, if hungry animals are offered feed with high concentrations of nitrate, or if nitrite is present in the feed, losses may be evident in a few hours. The onset of symptoms is usually rapid, and an animal may die within an hour after the first symptoms are noted. Cyanosis, rapid breathing, extreme respiratory movements, and nervousness are usually observed. Samples of blood obtained from animals in this condition are chocolate brown, and analysis of the blood shows a large percentage of methemoglobin. If the animal can be kept quiet and given methylene blue, recovery may be fairly rapid.

The livestock producer has a number of ways to guard against loss of livestock or decreased production from nitrate. These include a spot test to show the presence of nitrate and nitrite in feed and water, the addition of readily available carbohydrates to supplement the ration, limiting feed and water at each feeding, using methylene blue and glucose if an animal gets sick, management practices to reduce the amount of nitrate in forages, using properly protected surface impoundments for the water supply, and water management to prevent the concentration of nitrate by evaporation or freeze-out.

TABLE 20 The Toxicity to Animals of Nitrate and Nitrite in Drinking Water

Animal Tested	Concentration in Water (mg N/liter)	Effect	Reference
Nitrate			
Lambs	120[a]	Death	Pfander *et al.* (1957)
Lambs[b]	666	None	Seerley *et al.* (1965)
	1,000	16% of hemoglobin converted to methemoglobin	Seerley *et al.* (1965)
Pigs	300	None	Seerley *et al.* (1965)
Puppies	22–34	0.5–46% of hemoglobin converted to methemoglobin	Subbotin (1961)
	45–79	0.8–75% of hemoglobin converted to methemoglobin; 2 of 24 puppies died	Subbotin (1961)
Guinea pigs	1,130[c]	67% fetal loss	Sleight and Atallah (1968)
	60[c]	Abortion	Sinha (1969)
Nitrite			
Rats	$NaNO_2$, 100[c]	Growth inhibition, reduced life-span	Druckrey *et al.* (1963)
Pigs	100	None	Seerley *et al.* (1965)
Guinea pigs	240[c]	100% fetal loss	Sleight and Atallah (1968)

[a] Also nitrate in feed.
[b] Fed an alfalfa–corn diet.
[c] As mg/kg body weight.

PUBLIC HEALTH GUIDELINES

Because of the potential risk of methemoglobinemia to bottle-fed infants, public health agencies have established guidelines for acceptable concentrations of nitrate in drinking-water supplies. The acceptable tolerance level is usually set at 10 mg of nitrate nitrogen/liter of water (equivalent to 45 mg of nitrate/liter). In some communities, particularly in arid or semiarid regions, higher concentrations of nitrate occur naturally in underground water supplies; in these localities, the level is sometimes set at 20 mg of nitrate nitrogen/liter of water (90 mg of nitrate/liter), and the public is advised to be vigilant.

The occurrence of infant methemoglobinemia is usually associated with high levels of nitrate in shallow well water. In rural areas, this water can be directly contaminated by seepage from farm sites where nitrogen accumulation is especially great (e.g., barnyards, manure-disposal sites, and septic tank drainfields). As underground waters move away from such sites, the nitrate hazard is reduced because the nitrate concentration is diluted. These real but uncommon hazards are well known to residents of farming communities, public health workers, and veterinarians. Even though the means of guarding against contamination of well waters are known, continuing education and vigilance remain essential. Where sanitation standards are high, instances of infant methemoglobinemia have been extremely rare; in the western United States, for example, the last authenticated case was reported from Colorado in 1962 (Winton, 1970). It appears from the experience of public health authorities that presently accepted tolerance levels for nitrate in drinking waters are reasonable.

CARCINOGENIC, MUTAGENIC, AND TERATOGENIC EFFECTS OF NITROSAMINES

Dialkyl and related nitrosamines produce liver damage, hemorrhagic lung lesions, and convulsions and coma in rats in varying degrees (Heath and Magee, 1962). Dimethylnitrosamine was first studied with laboratory animals because of its acute or subacute effects on exposed personnel in industrial laboratories (Barnes and Magee, 1954). Some nitroso compounds have an acute toxicity so low that accurate LD_{50} data cannot be obtained (Magee and Barnes, 1962). These same studies show that acute toxic effects are not correlated with carcinogenic activity.

A series of experiments in the late 1950's and early 1960's [reviewed by Magee and Barnes (1967)] demonstrated that many nitrosamines are carcinogenic in various species of laboratory animals. In rats, they can induce malignant tumors at extremely low levels in the diet, such as 2 ppm (Terracini *et al.*, 1967). The nitrosamines vary in their carcinogenic potential and organ-specificity, and most produce tumors in rats–some with daily doses as low as 0.005–1.0 mg/kg of body weight (Druckrey *et al.*, 1967, 1969). One of the most striking characteristics of carcinogenic nitrosamines is that cancer can be induced after only a single oral LD_{50} dose of 30 mg/kg of body weight. Most animal species tested show tumors in the liver and, to a lesser extent, in the kidneys and bladder when challenged with dialkylnitrosamines. Asymmetrical nitrosamines cause tumors of the esophagus in the rat, whereas symmetrical dialkylnitrosamines predominantly induce liver tumors. Nitrosamine-induced tumors have also been produced experimentally in the kidneys, bladder, nasal sinuses, lungs and bronchi, the alimentary canal (esophagus, stomach, and small and large intestine), the nervous system, and the skin of rats (Magee and Barnes, 1967). Recent studies (DuPlessis *et al.*, 1969) on the high incidence of esophageal cancer among Bantu men in certain areas of the Transkei have raised the suspicion that this disease results from the dimethylnitrosamine found in the fruit of a solanaceous bush, the juice of which is used to curdle milk. In Zambia, geographic studies (McGlashan *et al.*, 1968) have linked cancer of the esophagus to the drinking of locally distilled spirits (kachasu) that contain dimethylnitrosamine at 1–3 ppm. This concentration would be carcinogenic in laboratory animals.

Simultaneous feeding of *N*-methylbenzylamine and nitrite to rats has been shown to produce esophageal tumors (Sander *et al.*, 1968). Indeed, one report indicates that simultaneous feeding of nitrite and secondary amines yields as many tumors as the feeding of the preformed nitrosamine (Greenblatt *et al.*, 1971). These observations raise the possibility that the simultaneous ingestion of nitrite (as from treated meat) and a secondary amine occurring naturally in the diet could present a human hazard. Nitrite could also be formed from nitrate, and secondary amines could be produced from tertiary amines by bacteria of the gastrointestinal tract. These possibilities cannot be assessed now.

Few scientific papers report experiments dealing with the possible effects of nitrosamines on the developing embryo and fetus—that is, teratogenic effects. However, it has been shown that *N*-

nitrosomethylurea and *N*-nitrosoethylurea are potent teratogens in rats, both inducing tumors in the progeny (Druckrey *et al.*, 1966). Administration of *N*-nitrosoethylurea prior to the twelfth day of gestation was found to produce no tumors, whereas administration during the last days of pregnancy gave the highest incidence of tumors. Other teratogenic effects were found by von Kreybig (1965), who gave pregnant rats a single dose of *N*-nitrosomethylurea intravenously. He found that many fetuses were killed and resorbed, and those surviving had many deformities.

Many carcinogenic nitrosamines are also mutagenic, and there is evidence for the production of chromosomal aberrations, as well as gene mutations. In this connection, *N'*-nitro-*N'*-nitrosomethylguanidine and *N*-nitrosomethylurea are perhaps the most potent chemical mutagens known (Magee and Barnes, 1967). The nitrosamines that require enzymic transformation before becoming active carcinogens are only active as mutagens under the same conditions (Malling, 1966). Such compounds as dimethylnitrosamine and diethylnitrosamine have been found to be active mutagens (Malling, 1971; Lijinsky, 1970).

The mechanism of mutagenesis by nitrosamines attracted some interest. Pasternak (1964) concluded that, after enzymic breakdown *in vivo*, nitrosamines exert their mutagenic action through alkylation by decomposition products. A similar conclusion was reached by Kihlman (1961) on the basis of studies of phenylmethylnitrosamine. The mechanism of mutagenesis by nitroso compounds was also considered by Fahmy *et al.* (1966), who compared the mutagenic activity of diethylnitrosamine and *N*-nitrosoethylurethane in *Drosophila*. Most authorities, however, feel that it is not possible to conclude that nitrosamines act as mutagens only by alkylation of the genetic material.

OTHER PATHOLOGICAL EFFECTS

A substantial body of evidence indicates that animals receiving nitrate or nitrite at subacute levels may suffer abnormalities. This is referred to as chronic toxicity. In some instances, the condition is a result of the conversion of a borderline nutrient deficiency to a gross deficiency. Whether the presence of subacute levels of nitrate or nitrite will induce these abnormalities will depend on many animal management factors; it has been established, however, that animals can be gradually adapted to high levels of nitrate (Clark *et al.*, 1970; Sokolowski, 1966) and nitrite (Holst *et al.*, 1961). This adaptation is probably associated with rumen microorganisms, the liver, and the kidney.

Nitrate is most likely to affect the very young, the old, the sick, and the poorly fed animal. Its toxicity is seen most frequently in cold weather, which can probably be attributed to increased feed requirements, limited availability of water (resulting in a large intake in a short time), and certain thyroid interrelationships.

THYROID

Nitrate can function as an antithyroid substance in the rat and other species (Bloomfield *et al.*, 1961; Lee *et al.*, 1970; Vought, 1966; Wyngaarden *et al.*, 1953), an effect probably resulting from "washing out" of iodine by nitrate (Bloomfield *et al.*, 1961). When nitrate is present, higher levels of iodine are needed to produce normal thyroid glands, and the dietary requirement increases from 35 to 200 ppb (Lee *et al.*, 1970). Field observations indicate that nitrate-induced thyroid pathology is most frequently seen in very cold environments.

NITRATE-VITAMIN A INTERRELATIONSHIP

High nitrate or nitrite levels in the feed and water of livestock sometimes lead to vitamin A deficiencies by destroying carotene or interfering with the utilization of vitamin A. Such deficiencies have been reported in survivors of acute toxicity. The oxides of nitrogen readily oxidize carotene and vitamin A, and nitrite destroys carotene more rapidly than nitrate (Emerick, 1963; Pugh and Garner, 1963). A number of studies in which the animals were given feed containing nitrate demonstrates that liver storage of vitamin A was reduced in many species (Bruggemann and Tiews, 1964; Hoar *et al.*, 1968; Mitchell *et al.*, 1965; O'Dell *et al.*, 1960; Wood *et al.*, 1967). Moreover, vitamin A specifically protects poultry from mortality when nitrate and nitrite are present in feed (Marrett and Sunde, 1968). It has also been reported that nitrate fed to pigs reduces serum vitamin A and E levels, although no signs of deficiency were evident (London *et al.*, 1967).

Some investigators have found that nitrate has little or no effect on vitamin A deficiencies. Thus, O'Donovan and Conway (1968) found that the vitamin A status of cattle was not affected by grazing in pastures treated with high levels of nitrate. Cunningham (1967) noted no effect with nitrite. Similarly, Michell *et al.* (1967) observed no influence of nitrate on preintestinal destruction of vitamin A, and many experiments on cattle in New York disclosed little or no effect on vitamin A activity (Davison *et al.*, 1964). Similar results were obtained by Wallace *et al.* (1964).

REPRODUCTION AND LACTATION

Asbury (1963) found that abortion in cattle was associated with a methemoglobin level of 40 percent and a loss of tone in the uterus. This finding suggests that the fetus is less able to withstand the effects of methemoglobin than the mother cow. The placenta, however, does not concentrate nitrate. In field conditions, abortion occurs several days after exposure to nitrate or nitrite.

Beef cows fed nitrate-rich silage from corn grown in drought years became irritable and urinated excessively and some aborted, but the cows recovered when the silage was removed (Pfander *et al.*, 1964). Milk production may also be lowered when cows are fed nitrate (Stewart and Merilan, 1958). Although abortions and other signs identical to those observed in cattle grazing on weedy lowlands have been produced in cattle fed nitrate (Simon *et al.*, 1959), other trials with cattle have not produced these effects [see Crawford *et al.* (1966) for review]. On the other hand, nitrate-rich drought corn silage fed to sheep did not affect lambing performance or the birth weight of lambs, and ewes and lambs consuming wheat pasture heavily fertilized with nitrogen performed as well as control animals (Pfander *et al.*, 1964).

The possible injury to the fetus and the newborn rat when the mother consumes water containing 3 percent of sodium nitrite has been examined by Gruener and Shuval (1970). The newborn rats contained methemoglobin levels of 5–10 percent at birth and the mothers levels of about 45 percent. Within a few days, the suckling rats showed normal methemoglobin levels, whereas the mothers, who were still on nitrite intake, contained high levels. Although there were no differences in birth weights and litter size between the test and control progeny, the weight gain of the progeny of the mothers ingesting nitrite was slower than that of the controls.

HYPERTENSION

The view that nitrite could have an effect on cardiac function in man was advanced many years ago. For example, in 1917, Mathews concluded that nitrite causes a fall in blood pressure. The heart was accelerated because of a depression of the vagus center as an indirect result of the fall in blood pressure. These conclusions were largely related to organic nitrites. The studies of Weiss *et al.* (1937), in which a sudden dose of 36 mg of nitrite nitrogen was given to men, showed that increased heart rate, decreased blood pressure, and circulatory collapse could occur after the ingestion of inorganic nitrite. Similar observations have been made in sheep, in which 7.1 mg of nitrite nitrogen per kg of body weight was associated with definite vasodilation (Holtenius, 1957).

Morton (1971) examined a series of mortality and morbidity patterns in relation to a variety of factors, including altitude, water hardness, and nitrate levels in waters of Colorado. He reported a possible association between nitrate levels and an emerging (1950–1960) hypertension pattern in one river valley in Colorado. However, there were other associations, including one apparently between hypertension and altitude. The possible association between nitrate and hypertension needs further study to evaluate cause-and-effect relationships and to establish quantitative data.*

DIGESTIBILITY AND FEED INTAKE

Some concern has been expressed about the effect of nitrate on feed intake by animals and on ration digestibility. Thus, Stoker *et al.* (1961) reported reduced total digestible nutrients and nitrogen balance in cattle fed timothy from land receiving high rates of nitrogen fertilization. However, Harris *et al.* (1961) found no change in coefficients of digestibility in calves fed hay that had been fertilized with nitrogen prior to harvest.

NITROGEN OXIDES IN SILOS

Nitric oxide, nitrogen dioxide, and their polymers have been long recognized as constituting a public health hazard in certain commercial and industrial processes. Such oxides of nitrogen are also pro-

*Recent work reports changes in the electroencephalograms of rats given water containing nitrite at concentrations of 30 mg nitrite nitrogen per liter and higher. These changes largely disappeared in the rats on the 30 mg/liter level on transfer to nitrite-free water; however, they continued over a 4-month period in rats receiving higher concentrations. There was no suggestion that the overt behavior of the rats was altered. However, in parallel studies on mice, behavioral changes were reported in animals receiving 600 mg/liter of nitrite nitrogen but not at lower levels (N. Gruener and H. I. Shuval, 1971, and personal communication).

duced during the fermentation of plant tissues containing nitrate. These gases are released from silage and may cause the death of farm workers (silo filler's disease). The reports of Delaney *et al.* (1956) and Lieb *et al.* (1958) deal with seven cases of exposure and three deaths. Animals confined to areas near the base of upright silos may die from the exposure (Scaletti, 1963).

CONCLUSIONS

1. The recommended limits for nitrate concentrations in drinking water (i.e., those established by the U.S. Public Health Service in 1962) are based on reasonable evidence that waters within these limits are not a health hazard. The nitrate concentrations in most supplies of potable surface water are within the limits, but concentrations in many groundwater supplies used for drinking exceed the limits.
2. The presence of dense populations of animals may result in increased nitrate levels in the underlying soil. Later, in drought conditions, this nitrogen may enter feed.
3. Nitrate may accumulate in certain vegetable crops, such as spinach and beets. This accumulation is primarily a function of species, variety, and environmental stresses, but it is also influenced by available nitrogen in soil.
4. Nitrate in foods, whether of natural origin or added in processing, is not toxic to human adults, provided the quantity added falls within the limits set by the Food and Drug Administration.
5. There is no evidence that nitrate in canned baby food (spinach included) is hazardous when consumed by infants more than 3 months old.
6. Methemoglobinemia resulting from consumption of food and water is primarily a condition found in children, especially in very young infants. No cases resulting from consumption of nitrate-rich food and water have been reported in adults. Standards for adding nitrate and nitrite to foodstuffs and for the nitrate content of drinking water generally provide adequate safeguards in respect to methemoglobinemia. However, these standards may not ensure safety for very young infants in extreme cases; for example, they may not ensure safety if formulas are boiled extensively or if treated meats, such as wieners, are fed.
7. A possible problem remains in those communities where well

water sources with high nitrate levels are used. Educational programs directed to mothers and pediatricians in these communities have been eminently successful in the last decade in reducing the incidence of infant methemoglobinemia.

8. The consequences of undetected single or repeated episodes of methemoglobinemia in infants fed on formulas containing water with high concentrations of nitrate are unknown. Because of the insensitivity of the usual diagnostic techniques in this "subclinical" range and the small number of exposed infants, it is unlikely that conventional epidemiological studies aimed at detecting possible effects would be rewarding.

9. There remains a residual uncertainty as to the reality of reported, but as yet unconfirmed, adverse effects (slowing of motor reflexes) of methemoglobinemia in the so-called "subclinical" range.

10. On the basis of the available evidence, the quantity of nitrite permitted by the Food and Drug Administration in certain food products appears to be safe.

11. Nitrite in high concentrations may also appear in certain food products during storage as a result of nitrate reduction.

12. Under certain circumstances, including acid conditions, nitrite may react with amines to give rise to nitrosamines. Little is known about the occurrence or importance of nitrosamine formation in nature, in food or feeds, or in the gastrointestinal tract.

13. Exposure to some of the nitrosamines can cause carcinogenic, mutagenic, and teratogenic effects in experimental animals. No evidence exists that the foods consumed by humans and animals in this country contain nitrosamines at levels that might induce such effects. The implications of the potential for nitrosamine formation must be examined to give assurance that nitrosamines do not become a health hazard.

14. Forage plants may accumulate significant quantities of nitrate when grown under drought conditions on soils rich in nitrogen, and these plants may cause death of livestock. The extent of these losses depends on the level of nitrogen in the soil, the species of plants, the stage of maturity of the plants, the severity of the drought, the prior condition of the exposed animals, and the management practices of the livestock producer.

15. Readily available kits for detecting nitrate can be used in screening feed supplies and in selecting samples for quantitative laboratory analysis. Most livestock producers in affected areas are aware of the possibility of acute nitrate toxicity and of means of testing for nitrate.

RECOMMENDATIONS

1. Equipment for removing nitrogen from drinking water should be devised for use in homes and on farms. Among the possible methods are microbiological denitrification and anion-exchange resins.

2. The Public Health Service recommended limits for nitrate in drinking water should not be relaxed.

3. The nitrate level in selected foods and drinking water supplies should continue to be measured to detect concentrations possibly injurious to health.

4. Although there are no demonstrated hazards arising from the presence of nitrate in vegetables in the United States, prudence dictates that where vegetables are intended for use in baby foods, nitrogen fertilizers should be applied sparingly, as required for satisfactory crop production but without luxury absorption of nitrate by the plant. Educational programs should be designed to bring this information to vegetable growers.

5. Standards for processed meat foods should be reviewed to determine whether the advantages of nitrate–nitrite can be obtained with lesser amounts.

6. Educational programs telling mothers in "high nitrate" areas to use uncontaminated (i.e., distilled, bottled) water in preparing formulas should be intensified. The programs should explain that nitrate-rich foods, such as spinach and beets, should not be given to infants less than about 3 months old, because of their susceptibility to methemoglobinemia.

7. Appropriate state and federal agencies should sponsor research on methods of preparing and handling baby foods in the home so as to avoid potential hazards to infants from foods high in nitrate. Information about these methods should then be made available to mothers.

8. Educational programs should be designed to inform livestock producers of the need for testing forage and water for nitrate and nitrite in regions or circumstances where a potential hazard to livestock exists.

9. Public agencies and private industry should find substitutes for nitrate and nitrite used for preservation and for enhancing the color of processed meats.

10. Research should be supported to establish in the field, and possibly in controlled laboratory studies, whether low levels of

methemoglobinemia in humans have functional consequences, such as slowing of motor reflexes.

11. In future studies of methemoglobinemia, particular attention should be given to identifying special host factors that may increase susceptibility of infants to nitrate. More detailed analyses of water sources are desirable. These analyses should include nitrate and nitrite content, coliform density, and biological oxygen demand or dissolved oxygen.

12. Means for enhancing nitrite reduction in the rumen should be devised.

13. Methods for accurately measuring low levels of nitrosamines should be developed.

14. The levels of nitrosamines likely to produce effects in man should be determined, and whether such levels are found in foods should be established.

15. Whether nitrosamines are likely to be formed in man by ingestion of nitrite- or nitrate-containing foods should be determined.

16. The possibility that nitrosamines are formed in soil and natural bodies of water in which nitrate or nitrite appears in significant amounts should be investigated.

Eutrophication

The enrichment of waters with nutrients is referred to as eutrophication. Although this process occurs under natural conditions, the activities of man in altering the landscape by urbanization and agricultural and industrial development, accompanied by the discharge of wastes, frequently localize and intensify the quantities of nutrients going into lakes, streams, and estuaries.

Eutrophication may have beneficial results by increasing productivity in receiving bodies of water, but in many locations throughout the world the effects of eutrophication are undesirable. An excess of nutrients often results in excessive growth of algae and larger aquatic plants, which interfere with the use of water for recreation, increase the cost of filtration of water for domestic and industrial purposes, reduce or fully consume the oxygen resource in the deeper or hypolimnetic waters, adversely affect aesthetic values, and impart taste and odor to the water. Furthermore, excessive growth of larger aquatic plants along waterways hinders drainage, and enhances the probability of flooding.

During the past decade, the scientific literature on the problems of eutrophication has been extensive. The subject is fully treated in *Eutrophication: Causes, Consequences, Correctives* (National Academy of Sciences, 1969) and in reports by Lee (1970b), Stewart and Rohlich (1967), and Vollenweider (1968).

Phosphorus and nitrogen, of the elements essential for the growth of phytoplankton and macrophytes, have been given particular attention; and of the forms of nitrogen in water, both ammonium and nitrate are available for the nutrition of aquatic plants. Nitrate concentrations in lake waters range from a few micrograms to several thousand micrograms per liter. In considering the nitrogen budgets of lakes and estuaries, however, it is necessary to deal with the transformations in nitrification, denitrification, and nitrogen fixation in the nitrogen cycle.

As noted earlier (page 6), a precise national nitrogen budget is not available. Consequently, the amount of nitrogen reaching surface waters and possibly having an influence on eutrophication is unknown except in those locations where nutrient budgets have been made.

With few exceptions, present data do not permit adequate descriptions of the quality of the water in lakes in this country. Published inventories giving lake name, size, location and—in most cases—depth are available for about 86,000 lakes, but data on water quality are generally lacking or fragmentary. Preliminary results of a survey being conducted at the University of Wisconsin have identified about 425 lakes in this country that have deteriorated to such an extent that rehabilitation is desirable (Ketelle and Uttormark, 1971).

Although phosphorus has been given particular attention in efforts to control eutrophication in lakes, Ryther and Dunstan (1971) point out that it is nitrogen that limits and controls algal growth and eutrophication in coastal waters and estuaries.

ANALYTICAL TECHNIQUES

Continuous evaluation and improvement of analytical techniques are required for effective control of water pollution (American Chemical Society, 1969). In particular, sampling techniques, sample preservation during transport and storage, and analytical methods should be carefully examined.

Sampling techniques are perhaps the most critical need, because of the importance of obtaining representative samples. Point sources such as wells, drains, and effluent discharges cannot always be characterized adequately by a single dip or grab sample. The concentration of nitrogenous compounds in such sources may vary considerably with time, and adequate samples should be obtained to measure this variation. Variability of lakes or streams with location may be assessed from analysis of samples collected at different points, consideration being given to variability with depth caused by thermal stratification and incomplete mixing. Variability with time is often related to stream flow, but the relationship is not simple: Concentrations of nitrate may increase or decrease depending on watershed characteristics. Other forms of nitrogen may remain constant or decrease (Frink, 1971). Thus, sampling in proportion to flow is often required, the ultimate instrument for this being a continuous composite sampler equipped for automatic proportional sampling according to flow. Clearly, the cost of sampling methods varies widely, and consideration should be given to the final use of the data before investing in costly sampling devices. As one example of the difficulty of measurement, where stream flow volumes and nutrient concentrations vary inversely, greater masses (in kilograms) of nutrients may be transported during high-flow periods even though the nutrient concentration may be more dilute. However, concentrations of streamborne nutrients may actually be high or low during high-flow periods (Frink, 1971). For streams having low concentrations during low-flow periods, less frequent sampling may be required during the low-flow periods whenever estimates of total nutrient removal by the stream are desired.

Once collected, samples must be preserved during transport and storage until analyses can be completed. Preserving the samples is of minor importance, however, if they are obtained solely to determine total nitrogen and if denitrification does not take place.

Distribution of nitrogen among the nitrate, nitrite, ammonium, and organic forms is particularly affected by microbial activity. Various microbial inhibitors have been used, but present evidence suggests that storage just above freezing may be preferable to the use of inhibitors. Again, the ultimate use of the data should be considered. Where health hazards are involved, the widely differing effects of the various nitrogenous compounds make it imperative that changes in sample composition be prevented.

Since legal action may require analyses of water, there is an obvious need for "standard" methods that may be used by any laboratory. Unfortunately, the acceptance of these methods as standard tends to stifle the search for better ones. Thus, the standard method for organic nitrogen remains virtually as described by Kjeldahl in 1883. While present methods for nitrate, nitrite, ammonium, and organic nitrogen are adequate for many purposes, they must be evaluated and improved where necessary.

MONITORING AND SURVEILLANCE

Monitoring and surveillance are necessary in any program to control water pollution. They make it possible to establish base-line data and observe changes with time and thus determine whether specific activities are polluting or cleansing the water. However, monitoring is expensive and, with automated methods of analysis, may produce data faster than they can be summarized and interpreted. As a case in point, the U.S. Geological Survey has 15-yr records for 100 watersheds throughout the country, yet water quality has not been related to land use. Despite this backlog of data, the Federal Water Quality Administration has estimated that at least 10,000 water-quality monitoring stations will be required by 1975. In addition, the states, municipalities, watershed authorities, and other local agencies are under constant pressure to monitor water quality.

This conflict between the ability to collect water-quality data, at considerable expense, and the inability to interpret such data must be resolved. A high priority should be given to evaluating existing data of the U.S. Geological Survey so that water quality may be related to land use. Archetypes among watersheds may likely be identified, and consideration should then be given to restricting future monitoring to such watersheds. It is also important to compare the value of monitoring point sources, such as municipal effluents, with the drawbacks of monitoring diffuse sources, such as land runoff. Point sources are easily identified and sampled.

Sufficient data have already been obtained to show that the concentration of nitrogen in many lakes and streams exceeds the minimum considered necessary for the growth of weeds and algae. Thus, rather than continuing to collect data to document the problem, it seems appropriate to direct more of our efforts toward reducing the nitrogen concentrations in waterways by minimizing nitrogen losses from all sources.

PREVENTION AND CONTROL

POINT SOURCES

The point sources of nitrogen are municipal and industrial waste waters, septic tanks, and feedlot discharges. Because estimates of quantities from these sources vary, each situation must be considered individually. Only through carefully conducted nutrient surveys at a particular location can decisions be reached as to the feasibility of control measures that might be adopted.

Municipal and Industrial Wastes Municipal and industrial waste waters are the point sources most suitable for application of nutrient control processes. Several methods have been developed for removing both phosphorus and nitrogen. Although these systems are designed principally to remove a single nutrient, they also reduce the amounts of other waste constituents. This must be kept in mind when assessing the improvements that result from nutrient removal.

Waste waters often contain a high initial content of organic nitrogen, which is subsequently converted to ammonium as a result of microbial activity. Aerobic treatment results in the oxidation of ammonium nitrogen to nitrite and nitrate. Under subsequent anaerobic conditions, denitrification may occur, resulting in the formation of free nitrogen gas.

Treatment to remove nitrogen from waste water effluents has been focused on a few of the forms of this element. Several basic processes have been reviewed by Eliassen and Tchobanoglous (1969), Farrell (1969), and Vollenweider (1968). A comparison of different methods is presented in Table 21. The following processes for removing nitrogen from water are considered: microbial assimilation, microbial denitrification, ammonia stripping, ion exchange, harvesting of algae, demineralization, and land application.

Removal of inorganic nitrogen by microbial assimilation is accomplished by providing a proper carbon-to-nitrogen ratio so that nitrogen is assimilated and organically bound in the microbial cells. Removal of the cellular mass by subsequent clarification results in removal of nitrogen from the system. The amount of nitrogen removed depends on the amount of growth attained, which in turn depends on the amount of food available. If the waste water is properly balanced, it is theoretically possible to convert all soluble forms of nitrogen into organic forms incorporated in microbial cells. In practice, the complete conversion of nitrogen to microbial protoplasm requires the addition of organic compounds as sources of nutrients and energy. Without this fortification, it has been observed that only 30–40 percent of the nitrogen in domestic sewage is converted to organic nitrogen by conventional secondary biological treatment processes.

To obtain nitrogen removal by microbial denitrification, a two-step

TABLE 21 Comparison of Nutrient Removal Processes for Domestic Waste[a]

Method	Removal Efficiency (%)		Estimated Cost (dollars per million gal)
	Nitrogen	Phosphorus	
Ammonia stripping	80–98	–	9–25
Denitrification	60–95	–	25–30
Algae harvesting	50–90	Varies	20–35
Conventional biological treatment	30–50	10–30	30–100
Ion exchange	80–92	86–98	170–300
Electrochemical treatment	80–85	80–85	4–8[b]
Electrodialysis	30–50	30–50	100–250
Reverse osmosis	65–95	65–95	250–400
Distillation	90–98	90–98	400–1000
Land application	Varies	60–90	75–150
Modified activated sludge	30–50	60–80	30–100
Chemical precipitation	–	88–95	10–70
Chemical precipitation with filtration	–	95–98	70–90
Sorption	–	90–98	40–70

[a]Source: Eliassen and Tchobanoglous (1969).
[b]For power only.

process is required: First, aerobic treatment allows for the conversion of ammonium and organic nitrogen to nitrate; second, anaerobic treatment favors the denitrifying bacteria that reduce the nitrate to elemental nitrogen gas. The anaerobic phase of this process usually requires supplemental organic carbon to ensure complete denitrification. The organic carbon sources that may be used include methanol, ethanol, acetone, and acetic acid. A major advantage of this process is that handling and disposing of excess sludge are not serious problems (Wuhrmann, 1964). The disadvantages include some nitrite accumulation and usually incomplete oxidation of the carbon source.

Ammonia stripping has also attracted attention. In most activated sludge plants, the nitrogen remaining in the final effluent is chiefly in the ammonium form. These ammonium ions exist in equilibrium with ammonia, and at a pH of about 11 most of the ammonium ions are converted to ammonia gas, which can subsequently be removed by air stripping. A drawback of the ammonia stripping process, as noted in studies by Nesselson (1954) and Kuhn (1956), is the high solubility of ammonia in water, which necessitates a high air-to-water ratio. About 450 ft^3 of air per gallon of sewage is required to achieve 95 percent removal (Eliassen and Tchobanoglous, 1969). Stripping towers having countercurrent low-pressure airflow can be used to provide the necessary air–solution interchange surface. Cold weather performance is often poor because the waste water is cooled by contact with the air and the solubility of ammonia consequently is increased. Maintenance costs in ammonia stripping may be high because of calcium carbonate deposits on the equipment. Also, before ammonia stripping is begun, consideration should be given to the possibility that the ammonia volatilized will return to the earth in the vicinity of the stripping operation.

Nitrate ions can be removed by contact with appropriate anion exchange resins, and ammonium ions can be removed by cation exchange. However, both systems have had problems because of fouling by organic materials and regeneration costs.

Nitrogen may be removed from waste water by harvesting algae grown in specially designed shallow ponds containing this water. Such removal depends on the transformation of nitrogen to algal cells and is analogous to the microbial assimilation process described above. To remove appreciable amounts of nitrogen by this method, it may be necessary to supplement the raw waste water with carbon dioxide and a carbon source, such as methanol. Eliassen and Tchobanoglous (1969) have predicted that 40–90 percent of the nitrogen can be removed by this process. Major disadvantages include the large land requirements, operational problems due to varying climatic conditions, difficulties in harvesting the algae, and the necessity of disposing of the algal sludge. It may be feasible, however, to incorporate the harvested algae into some type of high-protein animal feed, which would partially alleviate the disposal problem, provided the animal wastes are suitably recycled.

Demineralization processes can also be utilized for removing nitrogen compounds. The processes include electrolysis, reverse osmosis, and distillation. Since these processes are not selective, salts in addition to those of nitrogen will also be removed.

Chemical precipitation and membrane clogging have limited the usefulness of electrolysis for treating waste water; thus, salts with low solubilities may precipitate on the membrane surfaces, and colloidal organic matter, tending to collect on the membranes, may cause physical damage. Reverse osmosis separates water and ions by forcing the water from a solution having a high concentration of salts to one of lower concentration through cellulose acetate or other suitable membranes under pressures as high as 750 psi (St. Amant and Beck, 1970). The nitrate ion is one of the ions that is not removed completely. This fact limits the usefulness of the method. Other problems include concentration polarization, membrane fouling, and passage of certain ions through the membrane. Distillation involves heating of waste water to drive off relatively pure water vapor, which is subsequently condensed and collected, and treatment of distillates would be required to remove volatile substances.

Some waste waters may be treated by applying them to the land, either by ridge-and-furrow or spray irrigation. This technique is appealing because benefits from the waste water may be derived in some instances by utilizing the nutrients to produce usable crops. The major disadvantages are the large amounts of land required and the problems associated with operation in cold weather. The removal of nitrogen compounds in such a soil system is affected by physical adsorption on the various soil constituents and by microbial action; however, when the amounts of waste water and the nitrogen therein are more than necessary for crop growth, water draining from the root zone and moving to the groundwater will transport high concentrations of nitrate (Bouwer *et al.*, 1971). Therefore, this technique is of limited value for high-nitrate waters where groundwater contamination is a possibility.

Feedlots A variety of processes have been investigated for treating and disposing of wastes from animal feedlots. These include anaerobic and aerobic lagoons, oxidation ditches, composting, land application, chemical treatment, incineration, dehydration, and coprophagy. Detailed discussions on management of animal wastes are available in the proceedings of conferences at Cornell University, the University of Wisconsin, and Iowa State University in 1969.

DIFFUSE SOURCES

Agricultural Sources Losses of nitrogen from farm sites may be minimized by applying nitrogen to the growing crop, by tailoring fertilizer rates to crop use, and by selecting crop varieties to scavenge the soil for nitrogen and convert it into useful protein (Frink, 1971). Much animal waste is produced as an integral part of agricultural production. (An example would be the waste from the 7 million dairy cows in northeastern and north central United States.) Animal wastes should be returned to the soil with the same considerations for timing and rate of application as are given to other nitrogenous fertilizers. In irrigated areas, nitrogen in the irrigation water is frequently a major item in the nitrogen budget of the crop; indeed, studies have shown that the quality of irrigation water may actually be improved by cycling it through a crop.

Other measures–including slow-release chemical fertilizers, nitrification inhibitors, and more extensive use of winter cover crops–have been suggested and should be given full consideration (Aldrich, 1970). Despite these improved agronomic practices, however, there is evidence that some of the nitrogen applied to agricultural soil in excess of the amounts used by the crops is lost to waterways.

It is not known whether the nitrogen needs of row crops grown close to the maximum yield potential of the environment can be met without some nitrate leaching below the root zone. Plants may not be able to absorb all the nitrate from solution in the lower root zone without some decrease in yield. Reducing the fertility level of soils to lower the nitrate concentration in ground or surface water could seriously reduce our food supply, result in substantial increases in food prices, and greatly increase the acreage needed to satisfy our food and fiber requirements. Moreover, fertilizer restrictions would probably require expansion of intensive agriculture onto marginal land, which would increase erosion and, in some localities, might even result in nitrate percolation at a higher rate. More information is obviously needed on how much nitrate is leached under our present systems of high crop production and how much is lost by denitrification before major changes in agricultural practices and programs can be recommended. Nevertheless, farmers should be encouraged to manage their nitrogen programs efficiently and to expand the use of practices known to minimize erosion and runoff into lakes and streams. Fall or winter applications of nitrogen fertilizer or nitrogen-containing animal wastes appear to be especially hazardous practices in humid areas where subfreezing temperatures occur. Other practices (for example, split applications of fertilizer) can be adopted in those areas–practices that would reduce nitrate losses yet meet the needs of high-yielding crops. In some areas the changes would involve only a small cost.

Urban Sources Urban sources of nitrogen include fertilizer used on lawns, native soil organic nitrogen converted to inorganic forms, leachate from waste disposal in dumps or sanitary landfill sites, and nitric oxide and nitrogen dioxide gases from automobile exhausts and other combustion sources. Much remains to be determined with respect to the environmental impact of these sources, although measurements of urban runoff indicate that the total contribution of urban nitrogen to receiving waters may be considerable.

In communities with separate storm and sanitary sewers, storm drainage water might be used for irrigation. Contributions of nitrogen to waterways from municipal waste-disposal sites should be investigated, and means should be devised for reducing these contributions where significant. Studies by Robinson and Robbins (1968) indicate that 0.1 lb of nitrogen dioxide is emitted for every gallon of gasoline consumed.

Natural Sources Losses of nitrogen from all other categories of land use are included here as losses from natural sources, which include mineralization of native soil organic nitrogen, biological nitrogen fixation, geologic deposits of nitrate, and nitrogen in rainfall. In soils of the Midwest, which are high in native fertility, the mineralization of soil organic matter is a major source of nitrogen, and farmers have long known of this source of free fertilizer. In the humid East, native soil fertility is low, and losses from this source are much less. Aside from reducing surface erosion, it appears that little can be done to reduce losses from natural sources.

ANTICIPATED CONSEQUENCES AND COSTS OF CONTROL

A reduction in the yield of nitrogen to waterways would probably result in reduced growth of weeds and algae, a reduction most evident in waters where nitrogen is the element involved in limiting algal growth. The theory that a single nutrient may be limiting to algal development in lakes is not well established; however, a reduction of nitrogen supplied can be expected to reduce growth to some extent even in waters where other nutrients are limiting. Some rooted aquatic weeds seem capable of obtaining adequate phosphorus from enriched sediments although responding to nitrogen added to water. In addition, efforts to remove phosphorus from waterways may favor species capable of growing at lower phosphorus concentrations. Under these circumstances, a reduction in nitrogen may well cause a reduction in biomass. Because of the many uncontrollable sources of nitrogen in the environment–including nitrogen fixation by blue-green algae, waterfowl droppings, precipitation, and bodies of insects–it is unlikely that a reduction in nitrogen yield to waterways would eliminate weeds and algae; however, a reduction in growth in many waters can be expected.

The costs of control vary widely, and, in view of the reduction in growth of weeds and algae anticipated in some cases, only modest expenditures may be justified initially.

CRITERIA AND STANDARDS

Algae, like other plants, can adapt to a wide range of physiological conditions, including nutrient supply. It is difficult, therefore, to establish standards for nitrogen concentrations in water that will prevent or allow algal growth. Moreover, a change of conditions may favor the appearance of a species well suited to the new circumstances. Nevertheless, it does appear that a concentration of inorganic nitrogen of 0.3 mg/liter is a reasonable estimate of the limiting concentration for this nutrient (Sawyer *et al.*, 1945; Vollenweider, 1968). This figure is considered a first approximation in the establishment of water-quality standards for preventing eutrophication. The fact that it is considerably less than the current recommended limit for nitrate nitrogen in drinking water, 10 mg/liter, should be noted.

As Fox (1970) points out, however, the standards approach is misleading in several ways. First, adequate technical information is often not available. Second, the value criteria of those affected should be considered, but these criteria are difficult to ascertain. Third, the establishment of a standard, although comforting to many, implies a clear-cut line between good and bad that does not exist. Finally, establishment of a standard implies that it will be met, but frequently a standard is merely a basis for negotiation. Thus, Fox (1970) suggests that public policy should recognize that the process of achieving desired levels of quality will be an incremental one. The critical concentration of nitrogen that is estimated to cause eutrophication should probably be viewed as a target, rather than as an unenforceable standard for some waters.

REGULATORY AND LEGISLATIVE NEEDS

Considerable regulation of point sources of nutrients already exists in the form of various "clean water" laws. However, few of the laws deal directly with the problem of nitrogen, being limited in general to regulation of organic content and bacterial numbers in municipal effluent. Although some states are attempting to set standards for permissible levels of various nitrogenous compounds in surface waters and in municipal effluent, others (for example, Connecticut) may specify that wastes be treated to the extent that is technologically feasible. The latter approach allows for changes in standards and criteria as new information becomes available.

Regulation of diffuse sources requires a somewhat different approach, since no convenient measure of nutrient discharge is available. In large part, proposed regulations deal primarily with agricultural operations, because these are believed to be a major source of nitrogen. Where agricultural operations are judged responsible for water pollution, some legislative action may be necessary. Because of the many kinds of agriculture in the United States, such regulation should be considered at the state rather than the federal level. For example, states might enact legislation prohibiting the spreading of manure on frozen ground, preventing fall fertilization for spring-sown crops, and favoring the growing of winter cover crops.

Some form of financial support may be needed to encourage adoption of new or revised agricultural practices. The new Rural Environmental Assistance Program of the U.S. Department of Agriculture has as a primary objective the encouragement of practices to "prevent or abate agriculture-related pollution of water, land, and air for community benefit and the general public good." About $144 million was appropriated for this project in 1970–1971 to be used on a cost-sharing basis "where a partnership between the public and farmers to benefit both is clearly proper and fitting." If this money is wisely spent, and if supervision is at the state level, it may provide the initiative necessary for improving agricultural practices. Careful attention should be given to the effectiveness of this program and to the possible need for increased funding.

CONCLUSIONS

1. The role of nitrogen in eutrophication is not completely understood, and priority should be given to continued studies of typical eutrophic lakes in order to define their chemical and biological responses to increased nutrient inputs.

2. Nitrogen concentrations in many lakes and streams exceed the minimum considered necessary for excessive growth of algae and aquatic weeds.

3. Freedom from eutrophication cannot be ensured by establishing absolute standards for concentrations of nitrogen compounds, including nitrate. Any amount of fixed nitrogen entering a stream or lake raises its potential for supporting plant life, but because each body of water is distinctive as to depth, climate, and rates of water turnover, any set of regulatory conditions set up for one may fail to control eutrophication in another.

4. Nevertheless, there should be no decrease in efforts to control eutrophication. These efforts should be made with the assumptions that people will continue to live where they now live, that their food requirements will not decline, and that their waste-disposal needs and their desire for environmental amenities will not diminish.

5. Because of the great variability in climate and in the needs of crops, legislation concerned with nitrate in soil leachates, if deemed advisable, should provide for regulation on a regional or watershed basis. Sources of community water supply are already subject to regulation by water pollution control boards.

6. In many regions it is not clear how nitrogenous organic wastes can be disposed of without adding to the eutrophication of surface water and lowering the quality of underground water. Sanitary landfills, sewage-sludge disposal sites, and accumulations of animal manure are all potential sources of nitrate. The only certain method for permanently disposing of nitrogen in such materials is one in which nitrogen is chemically or biologically oxidized to nitrate, which is reduced to nitrogen gas by denitrification.

7. The quantity of nitrogen that human and animal wastes contribute to water may be reduced by returning the wastes to the soil, which would reduce the need for increased fixation of atmospheric nitrogen. Where land is not available, wastes may have to be disposed of in other ways, but much greater emphasis should be placed on recycling nitrogen through the food chain.

8. Nitrogen can be removed from municipal and industrial waste waters by advanced treatment methods.

9. Wastes from animal feedlots can also be disposed of by advanced waste methods, but the costs are apt to be high.

10. Methods for controlling the movement of nitrogen from septic tanks into water are not available. Present alternatives include the installation of sewers where feasible and the treatment of groundwater after contamination from this source.

11. Losses of nitrogen from agricultural sources may be minimized by applying the nitrogen to the growing crop, rather than to bare soil; by tailoring fertilizer rates to crop need; and by using slow-release fertilizers, nitrification inhibitors, and winter cover crops.

12. Little can be done to reduce losses of nitrogen from diffuse nonagricultural sources, other than to reduce surface erosion. A crop can be planted to utilize these sources; but, unless it is harvested and removed (as with a timber crop), net removal of nitrogen from the site by the crop is limited.

13. Where dairy animal wastes are part of a cropping scheme, they should be returned to the soil with due consideration for timing and rate of application, as with other nitrogenous fertilizers.

14. Nitrogen in urban drainage arises partly from emissions of nitrogen oxides from motor vehicles and high-temperature combustion, and these sources can perhaps be controlled. Animal droppings in the urban areas are also significant nitrogen sources. Where storm and sewer drains are separate, storm drainage might be stored and used for irrigation or treated for removal of nitrogen.

15. Little is known of the factors governing the release of nutrients from sediments in lakes and streams or of the contributions of wildlife to the eutrophication process.

16. Monitoring of the nitrogen contribution from all diffuse sources is expensive and does not reveal means for controlling the nitrogen release.

RECOMMENDATIONS

1. Further information on archetypes among eutrophic lakes should be provided to improve understanding of their chemical and biological responses to increased nutrient loading. This need arises because of the lingering uncertainty regarding the role of nitrogen in eutrophication.

2. Methods should be developed for reducing the input of nitrogen, from both point and diffuse sources, into bodies of water that may support an unwanted algal bloom if the nitrogen concentration were not reduced. Despite the uncertainty about the importance of this element in many instances of eutrophication, it is clear that increased nitrogen in many waters has promoted the growth of algae and rooted plants.

Summary

Nitrogen is essential for all living things, and it has often been the major limiting factor in crop production throughout the centuries. However, with the development of the Haber process to synthesize ammonia from the air, the chemical industry—at least in developed countries—has been able to meet the ever-increasing requirement for this nutrient. Nitrogen, in the form of nitrate and nitrite, has also been used widely for many years as a preservative and color fixative in meat products.

Some forms of nitrogen can be toxic to animals. Methemoglobinemia in infants has been related to a high level of nitrate in drinking water. The possibility that some forms of nitrogen may combine to form the extremely carcinogenic class of compounds called nitrosamines is particularly disturbing. Nitrogen also enhances the growth of aquatic vegetation, and excessive growth affects water quality and use.

The need for large quantities of fertilizer nitrogen to feed our increasing population, which is increasing its per capita consumption of animal protein, must be met by the fertilizer industry. The present level of animal protein consumption in the United States requires about 6 of the 7.5 million tons of fertilizer nitrogen produced annually. In specific local instances where use of fertilizer is excessive and has been shown to contribute to surface or groundwater pollution, limitations on use are justified. Fall application of nitrogen should be minimized in areas where excessive leaching or runoff can occur. Research concerned with use of fertilizer nitrogen should be intensified, both to increase efficiency of use and to minimize entry of nitrogen into waterways.

Feedlots or dairy operations should be encouraged to return animal manures to the land in concentrations that conform with generally accepted practices of use of nitrogen fertilizer.

Infant methemoglobinemia from nitrate or nitrite in food or water, although of concern in special circumstances in the past, is not a major or widespread problem at present. Since public health records have been kept, 350 cases have been reported in the United States. Most of these occurred between 1945 and 1950 and affected very young infants, and most were attributed to contaminated well water. The decline in cases is attributed to the education of physicians and mothers in the affected areas. It seems prudent, however, to maintain our present water standards for nitrate and nitrite and to strive for a minimum of nitrate in any foods given to infants under 3 months of age.

The possible presence of nitrosamines (organic compounds that are carcinogenic, teratogenic, and mutagenic) in meats, vegetables, and canned goods has aroused concern. Nitrate, nitrite, and secondary and tertiary amines are precursors of nitrosamines, which accounts for the concern over the nitrate and nitrite additives. Little is known about the levels of nitrosamines that may be hazardous to humans. Improvements are needed in the analytical procedures generally utilized for detecting and measuring nitrosamine concentrations at the levels that may be present in foods.

Nitrogen fertilization of farm lands, urban runoff, and municipal wastes appear to be factors in the eutrophication of some lakes and streams. However, some of the factors controlling eutrophication and algal growth and water plants are not well defined. Techniques should be developed to reduce nitrogen inputs into water bodies from both point and diffuse sources. A considerable amount of additional research information on nitrification and denitrification within soils, streams, and lakes must be developed.

The Committee found an appalling lack of information about the significance of the various sources and means of control of nitrogen in waterways; practical methods for reducing or increasing the quantity of nitrogen lost from the soil; the significance of nitrogen in limiting algal growth in lakes and rivers; the importance of nitrosamines in nature and in foods; the formation of nitrosamines in the gastrointestinal tract; and the "subclinical" hazards, if any, to man and animals arising from the consumption of water and food containing modest concentrations of nitrate.

Recommendations are made for expanded, imaginative research on all ramifications of nitrogen as a fertilizer, food constituent, food additive and preservative, and waste component of the farm and city. However, the Committee finds no evidence of danger to man, animals, or the global environment from present patterns of nitrogen fertilizer usage.

Persons Providing Information to the Committee

A. F. BARTSCH, Federal Water Quality Administration, U.S. Department of the Interior

E. J. BINKERD, Armour and Company

A. W. BOUCHAL, Colgate-Palmolive Company

R. D. CADLE, National Center for Atmospheric Research

HERRELL DEGRAFF, American Meat Institute

L. A. DOUGLAS, Rutgers University-The State University of New Jersey

W. H. DURUM, U.S. Geological Survey, U.S. Department of the Interior

LEO FRIEDMAN, Food and Drug Administration, U.S. Department of Health, Education, and Welfare

W. H. GARMAN, The Fertilizer Institute

M. A. GOLDBERG, Lever Brothers Company

W. L. HOLLIS, National Canners Association

S. JOHNSON, JR., Hampshire Chemical Division, W. R. Grace & Company

O. E. KOLARI, American Meat Institute Foundation

PAUL KOTIN, National Institute of Environmental Health Sciences, U.S. Department of Health, Education, and Welfare

W. C. KRUMREI, The Proctor and Gamble Company

L. J. MCCABE, Bureau of Water Hygiene, U.S. Department of Health, Education, and Welfare

P. L. MINOTTI, Cornell University

DEE MITCHELL, Monsanto Company
W. E. MORTON, University of Oregon Medical School
G. A. PURVIS, Gerber Products Company
ELMER ROBINSON, Stanford Research Institute
W. T. SAYERS, Federal Water Quality Administration, U.S. Department of the Interior
G. E. SMITH, University of Missouri
C. H. WADLEIGH, Agricultural Research Service, U.S. Department of Agriculture
HARRY WALTERS, Haughley Research Farms Limited (U.K.)

References

Aldrich, S. R. 1970. The influence of cropping patterns, soil management and fertilizer on nitrates, p. 152–169. *In* Proceedings of the Twelfth Sanitary Engineering Conference, College of Engineering, University of Illinois, Urbana.

Alexander, R. A., G. A. Barden, J. F. Hentges, Jr., J. T. McCall, and W. K. Robertson. 1961. Nutrient composition and digestibility of corn silage as affected by fertilizing rate and plant spacing. J. Dairy Sci. 44:975.

Allison, F. E. 1957. Nitrogen and soil fertility, p. 85–94. *In* Soil. The yearbook of agriculture 1957. U.S. Department of Agriculture. U.S. Government Printing Office, Washington, D.C.

Allison, F. E. 1965. Evaluation of incoming and outgoing processes that affect soil nitrogen, p. 573–606. *In* W. V. Bartholomew and F. E. Clark (ed.), Soil nitrogen. American Society of Agronomy, Madison, Wisconsin.

Allison, F. E. 1966. The fate of nitrogen applied to soils. Adv. Agron. 18:219–258.

Allison, F. E., E. M. Roller, and J. E. Adams. 1959. Soil fertility studies in lysimeters containing Lakeland sand. U.S. Dep. Agr. Tech. Bull. 1199. 62 p.

American Chemical Society. 1969. Cleaning our environment. The chemical basis for action. Washington, D.C. 249 p.

American Meat Institute Foundation. 1970. Current status of the use of nitrate and nitrite in the meat industry; the nitrate–nitrite–nitrosamine question. Prepared for presentation to the NAS–NRC Committee on Nitrate Accumulation (unpublished report). (Copy on file at Agriculture Board, NRC, Washington, D.C.)

Asbury, A. C. 1963. Sublethal effects of nitrate, p. 57–58. *In* Proc. Conference on Nitrate Accumulation and Toxicity, Cornell Univ. Mimeo 64-6. Department of Agronomy, Cornell University, Ithaca, New York.

Aussannaire, M., C. Joly, and A. Pohlmann. 1968. Methemoglobinemie acquise du nourrisson par eau de canalisation urbaine. Presse Med. 76:1723–1728.

Bailey, W. P. 1966. Methemoglobinemia–acute nitrate poisoning in infants. Second report. J. Am. Osteopath. Assoc. 66:431.

Barker, A. V., N. H. Peck, and G. E. MacDonald. 1971. Nitrate accumulation in vegetables. I. Spinach grown on upland soils. Agron. J. 63:126–129.

Barnes, J. M., and P. N. Magee. 1954. Some toxic properties of dimethylnitrosamines. Brit. J. Ind. Med. 11:167–174.

Barnett, A. J. G. 1953. The reduction of nitrate in mixtures of minced grass and water. J. Sci. Food Agr. 4:92–96.

Barrows, H. L., and V. J. Kilmer. 1963. Plant nutrient losses from soils by water erosion. Adv. Agron. 15:303–316.

Bartik, M. 1964. Certain quantitative relations of nitrate and nitrite metabolism in farm animals with special regard to origin and development of methemoglobinemia caused nitrates and the diagnosis of poisoning. I, II, III. Folia vet. Kosice. 8:83–93, 95–109, 111–112.

Beard, R. R., and G. A. Wertheim. 1967. Behavioral impairment associated with small doses of carbon monoxide. Am. J. Public Health 57:2012–2022.

Becker, M. 1967. Nitrat und Nitrit in der Tiernahrung. Qual. Planat. Mat. Veg. 15 (1):48–64.

Biggar, J. W., and R. B. Corey. 1969. Agricultural drainage and eutrophication, p. 404–445. *In* Eutrophication: Causes, consequences, correctives. NAS–NRC Publ. 1700. National Academy of Sciences, Washington, D.C.

Bloomfield, R. A., C. W. Welsch, G. B. Garner, and M. E. Muhrer. 1961. Effect of dietary nitrate on thyroid function. Science 134:1690.

Bodansky, O. 1951. Methemoglobinemia and methemoglobin-producing compounds. Pharmacol. Rev. 3:144.

Bormann, F. H., G. E. Likens, D. W. Fisher, and R. S. Pierce. 1968. Nutrient loss accelerated by clear-cutting of a forest ecosystem. Science 159:882–884.

Bouwer, H., J. C. Lance, and R. C. Rice. 1971. Nitrogen in soil and water. University of Guelph, Ontario. 169 p.

Bower, C. A., and L. V. Wilcox. 1969. Nitrate content of the upper Rio Grande as influenced by nitrogen fertilization of adjacent irrigated lands. Soil Sci. Soc. Am. Proc. 33:971–973.

Bradley, W. B., H. F. Eppson, and O. A. Beath. 1940. Livestock poisoning by oat hay and other plants containing nitrate. Univ. Wyo. Bull. 241 p.

Brown, J. R., and G. E. Smith. 1966. Soil fertilization and nitrate accumulation in vegetables. Agron. J. 58:209–212.

Brown, J. R., and G. E. Smith. 1967. Nitrate accumulation in vegetable crops as influenced by soil fertility practices. Univ. Mo. Agr. Exp. Sta. Res. Bull. 920.

Bruggemann, J., and J. Tiews. 1964. Effect of NO_2^- ions and different environmental temperatures on carotene metabolism of chickens. Intern. Z. Vitaminforsch. 34:233–240. Cited in Nutr. Abstr. 1965, 35:56–57, #376.

Buchman, D. T., R. L. Shirley, and G. B. Killinger. 1968. Nitrate, ammonia, and methemoglobin in sheep when fed millet containing different levels of molybdenum and copper. Soil Crop Sci. Soc. Fla. Proc. 28:209–215.

Burden, E. H. W. J. 1961. The toxicology of nitrates and nitrites with particular reference to the potability of water supplies. Analyst 86:429–433.

Carter, J. N., O. L. Bennett, and R. W. Pearson. 1967. Recovery of fertilizer nitrogen under field conditions using nitrogen-15. Soil Sci. Soc. Am. Proc. 31:50–56.

Case, A. A. 1957. Some aspects of nitrate intoxication in livestock. J. Am. Vet. Med. Assoc. 130:323–329.

Case, A. A. 1970. The health effects of nitrates in water. Nitrate and water supply: Source and control, p. 40–46, 61, 65. *In* Proceedings of the Twelfth Sanitary Engineering Conference, College of Engineering, University of Illinois, Urbana.

Clark, J. L., W. H. Pfander, R. A. Bloomfield, G. F. Krause, and G. B. Thompson. 1970. Nitrate containing rations for cattle supplemented with either urea or soybean meal. J. Anim. Sci. 31(5):961–966.

Clarke, E. G. C., and M. L. Clarke. 1967. Garner's veterinary toxicology. Williams & Wilkins, Baltimore, Maryland.

Committee on Nutrition, American Academy of Pediatrics. 1970. Infant methemoglobinemia. The role of dietary nitrate. Pediatrics 46:475–478.

Cooper, C. F. 1969. Nutrient output from managed forests, p. 446–463. *In* Eutrophication: Causes, consequences, correctives. NAS-NRC Publ. 1700. National Academy of Sciences, Washington, D.C.

Corey, R. B., A. D. Hasler, G. F. Lee, F. H. Schraufnagel, and T. L. Wirth. 1967. Excessive water fertilization. Report to the Water Subcommittee, Natural Resources Committee of State Agencies, Madison, Wisconsin.

Cornell University. 1969. Proceedings of conference on animal waste management (Syracuse, New York, January 13–15, 1969). Cornell University, Ithaca, New York. 414 p.

Crabtree, K. T. 1970. Nitrate variation in ground water. Technical Completion Report. OWRR B-044-Wis. Office of Water Resources Research.

Crawford, R. F., W. K. Kennedy, and K. L. Davison. 1966. Factors influencing the toxicity of forages that contain nitrate when fed to cattle. Cornell Vet. 56:3–17.

Cunningham, G. N. 1967. The effect of nitrite and hydroxylamine on the performance and vitamin A and carotene metabolism of ruminants. Diss. Abstr. (B) 27(11):3731B.

Davison, K. L., W. M. Hansel, L. Krook, K. McEntee, and M. J. Wright. 1964. Nitrate toxicity in dairy heifers. 1. Effects on reproduction, growth, lactation and vitamin A nutrition. J. Dairy Sci. 47:1065–1073.

Davison, K. L., K. McEntee, and M. J. Wright. 1965. Response in pregnant ewes fed forages containing various levels of nitrate. J. Dairy Sci. 48:968–977.

Delaney, L. T., H. W. Schmidt, and C. F. Stroebel. 1956. Silo-filler's disease. Proc. Staff Meet., Mayo Clinic 31(7):189–198.

Delwiche, C. C. 1970. The nitrogen cycle. Sci. Am. 223:137–146.

Dilz, K., and J. W. Woldendorp. 1960. Distribution and nitrogen balance of 15-N-labelled nitrate applied on grass sod, p. 150–152. *In* Proceedings of the 8th International Grassland Congress, Reading, England.

Diskalenko, A. P. 1968. Methemoglobinemia of water–nitrate origin in the Moldavian S.S.R. Hyg. Sanit. 33:32–37.

Diven, R. H., R. E. Reed, and W. J. Pistor. 1964. The physiology of nitrite poisoning in sheep. Ann. N.Y. Acad. Sci. 3:638–643.

Downs, E. F. 1950. Cyanosis of infants caused by high nitrate concentrations in rural water supplies. Bull. World Health Organ. 3:165–169.

Druckrey, H., D. Steinhoff, H. Beuthner, H. Schneider, and P. Klarney. 1963. Testing of nitrates for chronic toxicity in rats. Arzneimittel-Forsch. 13:320.

Druckrey, H., S. Ivankovic, and R. Preussmann. 1966. Teratogenic and carcinogenic effects in the offspring after single injection of ethylnitroso-urea to pregnant rats. Nature 210:1378–1379.

Druckrey, H., R. Preussmann, S. Ivankovic, and D. Schmahl. 1967. Organotrope carcinogene Wirkungen bei 65 Verschiedenen N-Nitroso-Verbindungen an BD-Ratten. Z. Krebsforsch. 69:103–201.

Druckrey, H., R. Preussmann, and S. Ivankovic. 1969. N-nitroso compounds in organotropic and transplacental carcinogenesis. Ann. N.Y. Acad. Sci. 163: 676–696.

DuPlessis, L. S., J. R. Nunn, and W. A. Roach. 1969. Carcinogen in a Transkeian Bantu food additive. Nature 222:1198–1199.

Eliassen, R., and G. Tchobanoglous. 1969. Removal of nitrogen and phosphorus from waste water. Environ. Sci. Tech. 3:6:536–541.

Emerick, R. J. 1963. Sublethal effects of nitrate, p. 45–46. *In* Proceedings of a Conference on Nitrate Accumulation and Toxicity. Cornell Univ. Mimeo 64-6. Department of Agronomy, Cornell University, Ithaca, New York.

Emerick, R. J., L. B. Embry, and R. W. Seerley. 1965. Rate of formation and reduction of nitrite-induced methemoglobin *in vitro* and *in vivo* as influenced by diet of sheep and age of swine. J. Anim. Sci. 24(1):221–230.

Ender, F., and L. Ceh. 1967. Second conference on tobacco research proceedings. Freiburg, Germany, p. 83–91.

Ender, F., and L. Ceh. 1968. Occurrence of nitrosamines in foodstuffs for human and animal consumption. Food Cosmet. Toxicol. 6:569–571.

Ender, F., G. Havre, A. Helgebostad, N. Koppang, R. Madsen, and L. Ceh. 1964. Isolation and identification of a hepatotoxic factor in herring meal produced from sodium nitrite preserved herring. Naturwissenschaften 51:637.

Ewing, M. C., and R. M. Mayon-White. 1951. Cyanosis in infancy from nitrates in drinking-water. Lancet 1:931–934.

Fahmy, O. G., M. J. Fahmy, J. Massasso, and M. Ondrej. 1966. Differential mutagenicity of the amine and amide derivatives of nitroso compounds in *Drosophila melanogaster*. Mutation Res. 3:201–217.

Farrell, J. B. 1969. Physical–chemical methods for nitrogen removal, p. 220–226. *In* Proceedings of nutrient removal and advanced waste treatment symposium. U.S. Dep. Interior, Fed. Water Poll. Control Admin., Ohio Basin Region.

Farrell, J. B. 1971. Nitrogen in industry. Paper presented at Symposium on Nitrogen in Soil and Water, Hespeler, Ontario, Canada. March 30–31.

Fassett, D. W. 1966. Nitrates and nitrites, p. 250–256. *In* Toxicants occurring naturally in foods. NAS-NRC Publ. 1354. National Academy of Sciences, Washington, D.C.

Fox, I. K. 1970. The use of standards in achieving appropriate levels of tolerance. Proc. Nat. Acad. Sci. U.S. 67:877–886.

Frink, C. R. 1967. Nutrient budget: Rational analysis of eutrophication in a Connecticut lake. Environ. Sci. Tech. 1:425–428.

Frink, C. R. 1969. Water pollution potential estimated from farm nutrient budgets. Agron. J. 61:550–553.

Frink, C. R. 1970. The nitrogen cycle of a dairy farm, p. 127–133. *In* Relationship of agriculture to soil and water pollution. Cornell University, Ithaca, New York.

Frink, C. R. 1971. Plant nutrients and water quality. Agr. Sci. Rev. 9(2):11–25.

Gallagher, J. J., and L. W. Westerstrom. 1970. Coal–bituminous and lignite, p. 309. *In* 1969 Bureau of Mines mineral yearbook. Bureau of Mines, U.S. Department of the Interior. U.S. Government Printing Office, Washington, D.C.

Gibson, Q. H. 1943. The reduction of methemoglobin by ascorbic acid. Biochem. J. 37:615–618.

Greenblatt, M., S. Mirvish, and B. T. So. 1971. Nitrosamine studies: Induction of lung adenomas by concurrent administration of sodium nitrite and secondary amines in Swiss mice. J. Nat. Cancer Inst. 46:1029–1034.

Greene, I., and E. P. Hiatt. 1954. Behavior of the nitrate ion in the dog. Am. J. Physiol. 176:463–467.

Gruener, N., and H. I. Shuval. 1970. Health aspects of nitrates in drinking water, p. 89–106. *In* H. I. Shuval (ed.) Developments in water quality research. Humphrey Science Pub., Ann Arbor.

Gruener, N., and H. I. Shuval. 1971. Report presented at annual meeting of American Public Health Association, Minneapolis, Minnesota. October 9–16.

Harmeson, R. H., F. W. Sollo, Jr., and T. E. Larson. 1971. The nitrate situation in Illinois. J. Am. Water Works Assoc. 63:303–310.

Harris, R. R., W. B. Anthony, J. G. Starling, and C. A. Brogden. 1961. The influence of nitrogen fertilization and harvest time on the nutritive value of pelleted coastal bermudagrass hay. J. Dairy Sci. 44:973.

Heath, D. F., and P. N. Magee. 1962. Toxic properties of dialkylnitrosamines and some related compounds. Brit. J. Ind. Med. 19:276–282.

Hoar, D. W., L. B. Embry, and R. J. Emerick. 1968. Nitrate and vitamin A interrelationships in sheep. J. Anim. Sci. 27(6):1727–1733.

Holst, W. O., L. M. Flynn, G. B. Garner, and W. H. Pfander. 1961. Dietary nitrate vs. sheep performance. J. Anim. Sci. 20:936, Abstr. #145.

Holtenius, P. 1957. Nitrite poisoning in sheep with special reference to the detoxification of nitrite in the rumen. Acta Agr. Scand. 7:113–163.

Hoover, S. R., and L. B. Jasewicz. 1967. Agricultural processing wastes: Magnitude of the problem, p. 187–203. *In* N. C. Brady [ed.] Agriculture and the quality of our environment. American Association for the Advancement of Science, Washington, D.C.

Hutchinson, G. L., and F. G. Viets, Jr. 1969. Nitrogen enrichment of surface water by absorption of ammonia volatilized from cattle feedlots. Science 166:514–515.

Iacovoni, P., F. De Benedictis, R. Cassoni, V. Lucisano, and G. Gambelli. 1968. Interference of reserpine with some sodium nitrite effects on rabbits. Gazz. Int. Med. Chir. 73(24, Pt. 3):5964–5968. *In* Chem. Abstr. 1970. 73:43656.

Jackson, W. A., J. S. Steel, and V. R. Boswell. 1967. Nitrates in edible vegetables and vegetable products. Proc. Am. Soc. Hort. Sci. 90:349–352.

Jaworski, N. A., and L. J. Hetling. 1970. Relative contributions of nutrients to the Potomac River basin from various sources, p. 134–146. *In* Relationship of agriculture to soil and water pollution. Cornell University, Ithaca, New York.

Jenny, H. 1933. Soil fertility losses under Missouri conditions. Mo. Agr. Exp. Sta. Bull. 324.

Jones, I. R., P. H. Weswig, J. F. Bone, M. A. Peters, and S. O. Alpan. 1966. Effect of high nitrate consumption on lactation and vitamin A nutrition of dairy cows. J. Dairy Sci. 49:491–499.

Kamm, L., G. G. McKeown, and D. M. Smith. 1965. New colorimetric method for the determination of the nitrate and nitrite content of baby food. J. Assoc. Off. Anal. Chem. 48:892–897.

Kaufmann, R. F. 1970. Hydrogeology of solid waste disposal sites in Madison, Wisconsin. USDI Tech. Report OWWR A-0180 Wis. Water Resources Center, University of Wisconsin, Madison. 361 p.

Ketelle, M. J., and P. D. Uttormark. 1971. Problem lakes in the United States. Tech. Report 16010EHR. Water Resources Center, University of Wisconsin, Madison.

Kihlman, B. A. 1961. Cytological effects of phenylnitrosamines. II. Radiomimetic effects. Radiat. Bot. 1:43–50.

Kilmer, V. J., O. E. Hays, and R. J. Muckenhirn. 1944. Plant nutrient and water losses from Fayette silt loam as measured by monolith lysimeters. J. Am. Soc. Agron. 36:249–263.

Knotek, Z., and P. Schmidt. 1964. Pathogenesis, incidence and possibilities of preventing alimentary nitrate methemoglobinemia in infants. Pediatrics 34:78–83.

Kravitz, H., L. D. Elegant, E. Kaiser, and B. M. Kagan. 1956. Methemoglobin values in premature and mature infants and children. AMA J. Dis. Child. 91:1–5.

Kuhn, P. A. 1956. Removal of ammonia nitrogen from sewage effluent. M.S. thesis. University of Wisconsin, Madison.

Larson, T. E., and L. Henley. 1966. Occurrence of nitrate in well waters. Final report. Project 65-05G. University of Illinois Water Resources Center, Urbana.

Lathwell, D. J., D. R. Bouldin, and W. S. Reid. 1970. Effects of nitrogen fertilizer applications in agriculture, p. 192–206. *In* Relationship of agriculture to soil and water pollution. Cornell University, Ithaca, New York.

Lee, C., R. Weiss, and D. J. Horvath. 1970. Effects of nitrogen fertilization on the thyroid function of rats fed 40% orchard grass diets. J. Nutr. 100(10): 1121–1126.

Lee, D. H. K. 1970a. Nitrates, nitrites, and methemoglobinemia. Environ. Res. 3:484–511.

Lee, G. F. 1970b. Eutrophication. Occasional Paper No. 2, University of Wisconsin, Water Resources Center, Madison.

Lewis, D. 1951a. The metabolism of nitrate and nitrite in the sheep. 1. The reduction of nitrate in the rumen of the sheep. Biochem. J. 48:175–180.

Lewis, D. 1951b. The metabolism of nitrate and nitrite in the sheep. 2. Hydrogen donators in nitrate reduction by rumen microorganisms *in vitro*. Biochem. J. 49:149–153.

Lieb, G. M. P., W. N. Davis, T. Brown, and M. McQuiggan. 1958. Chronic pulmonary insufficiency secondary to silo-filler's disease. Am. J. Med. 24:471–474.

Lijinsky, W. 1970. Chemistry and biology of nitrosamines. Environ. Mutagen Soc. Newsl. 3:35.

Lipman, J. G., and A. B. Conybeare. 1936. Preliminary note on the inventory and balance sheet of plant nutrients in the United States. N.J. Agr. Exp. Sta. Bull. 607.

London, W. T., W. Henderson, and R. F. Cross. 1967. An attempt to produce chronic nitrite toxicosis in swine. J. Am. Vet. Med. Assoc. 150:398–402.

Magee, P. N., and J. M. Barnes. 1962. Induction of kidney tumours in the rat with dimethylnitrosamine (*N*-nitrosodimethylamine). J. Pathol. Bacteriol. 84:19–31.

Magee, P. N., and J. M. Barnes. 1967. Carcinogenic nitroso compounds. Adv. Cancer Res. 10:163–246.

Malling, H. V. 1966. The mutagenicity of two important carcinogens: Dimethylnitrosamine and diethylnitrosamine in *Neurospora crassa.* Mutat. Res. 3:537–540.

Malling, H. V. 1971. Dimethylnitrosamine: Formation of mutagenic compounds by interaction with mouse liver microsomes. Mutat. Res. 13:425–429.

Marrett, L. E., and M. L. Sunde. 1968. The use of turkey poults and chickens as test animals for nitrate and nitrite toxicity. Poult. Sci. 47:511–519.

Marquardt, P., and L. Hedler. 1966. Über das Vorkommen von Nitrosaminen in Weizenmehl. Arzneimittel-Forsch. 16:778–779.

Mathews, A. P. 1917. The pharmacology of nitrates and nitrites, p. 497–538. *In* H. S. Grindley and W. S. MacNeal (ed.) Studies in nutrition. Lakeside Press, Chicago.

Mayo, N. S. 1895. Cattle poisoning by nitrate of potash. Kansas State Agr. Exp. Sta. Bull. p. 49.

McCarthy, P. L., J. H. Hem, D. Jenkins, G. F. Lee, J. J. Morgan, R. S. Robertson, R. W. Schmidt, J. M. Symons, and M. V. Trexler. 1967. Sources of nitrogen and phosphorus in water supplies. J. Am. Water Works Assoc. 59:344–366.

McGlashan, N. D., C. L. Walters, and A. E. M. McLean. 1968. Nitrosamine in African alcoholic spirits and oesophageal cancer. Lancet 2:1017.

Miale, J. B. 1967. Laboratory medicine hematology. C. V. Mosby, St. Louis. 531 p.

Miller, W. E., and J. C. Tash. 1967. Interim report. Upper Klamath Lake studies, Oregon. FWPCA Publ. WP-20-8. National Environmental Research Center, Corvallis, Oregon. 37 p.

Mitchell, G. E., C. O. Little, and T. R. Greathouse. 1965. Influence of nitrate and nitrite on carotene disappearance from the rat intestine. Life Sci. 4(3):385–390.

Mitchell, G. E., C. O. Little, and B. W. Hayes. 1967. Pre-intestinal destruction of vitamin A by ruminants fed nitrate. J. Anim. Sci. 26:827–829.

Miyazaki, A. 1967. Studies on the effects of nitrate in feed upon the performance of ruminants. 1. Effects of nitrate added to feed upon the gains and blood constituents of sheep. Jap. J. Zootech. Sci. 38:527–536; 1968, 39:20–26. Cited by Nutr. Abstr. 1968, 38(4):1412.

Mohler, K., and O. L. Mayrhofer. 1968. Nachweis und Bestimmung von Nitrosaminen in Lebensmitteln. Z. Lebensmittel-Unters. Forsch. 135:313–318.

Morton, W. E. 1971. Hypertension and drinking water constituents in Colorado. Am. J. Public Health 61:1371–1378.

Muenscher, W. C. 1961. Poisonous plants of the United States. The Macmillan Co., New York.

Mukherjee, S. K., and R. De. 1968. Response of potatoes to foliar application of nitrogen and phosphorus. Ind. J. Agr. Sci. 38:275–285.

National Academy of Sciences. 1969. Eutrophication: Causes, consequences, correctives. NAS-NRC Publ. 1700. National Academy of Sciences, Washington, D.C. 661 p.

Nelson, L. W. 1966. Nitrite toxicosis and the gastric ulcer complex in swine. Part I. Nitrite toxicosis. Diss. Abstr. 26:4140–4141.

Nesselson, E. 1954. Removal of inorganic nitrogen from sewage effluent. Ph.D. thesis. University of Wisconsin, Madison.

Newspaper Enterprise Association. 1969. The world almanac and book of facts, 1970 edition. Newspaper Enterprise Association, Inc., New York.

O'Dell, B. L., Z. Erek, L. Flynn, G. B. Garner, and M. E. Muhrer. 1960. Effects of nitrite containing rations in producing vitamin A and vitamin E deficiencies in rats. J. Anim. Sci. 19:1280.

O'Donovan, P. B., and A. Conway. 1968. Performance and vitamin A status of sheep grazing high nitrate pastures. J. Brit. Grassland Soc. 23(3):228–233.

Olson, O. E., and A. L. Moxon. 1942. Nitrate reduction in relation to oat poisoning. J. Am. Vet. Med. Assoc. 782:403–406.

Orgeron, J. D., J. D. Martin, C. T. Caraway, R. M. Martine, and G. H. Hauser. 1957. Methemoglobinemia from eating meat with high nitrite content. Public Health Rep. 72:189–193.

Palmer, A. W. 1902. Chemical survey of the water of Illinois. Report for years 1897–1902. Bull. 2. Illinois State Water Survey.

Parker, C. A. 1957. Non-symbiotic nitrogen fixing bacteria in soil. III. Total nitrogen changes in a field soil. J. Soil Sci. 8:48–59.

Pasternak, L. 1964. Untersuchungen über die mutagen Wirkung verschiedener Nitrosamin- und Nitrosamid-Verbindungen. Arzneimittel-Forsch. 14:802–804.

Pease, H. T. 1896. Poisoning of cattle by andropogon sorghum. Agr. Ledger 3:221–226.

Petukhov, N. I., and A. V. Ivanov. 1970. Investigation of certain psychophysiological reactions in children suffering from methemoglobinemia due to nitrates in water. Hyg. Sanit. 35:29–31.

Pfander, W. H., G. B. Garner, W. C. Ellis, and M. E. Muhrer. 1957. The etiology of nitrate poisoning in sheep. Mo. Agr. Exp. Sta. Res. Bull. No. 637.

Pfander, W. H., G. B. Thompson, G. B. Garner, L. M. Flynn, and A. A. Case. 1964. Chronic nitrate–nitrite toxicity from feed and water, p. 3–12. Mo. Agr. Exp. Sta. Spec. Rep. 38.

Phillips, W. E. J. 1971. Naturally occurring nitrate and nitrite in foods in relation to infant methemoglobinemia. Food Cosmet. Toxicol. 9:219–228.

Preul, H. C. 1967. Underground movement of nitrogen, p. 309–328. *In* Advances in water pollution research. Water Pollution Control Federation, Washington, D.C.

Pugh, D. L., and G. B. Garner. 1963. Reaction of carotene with nitrite solutions. J. Agr. Food Chem. 11:528–529.

Robinson, E., and R. C. Robbins. 1968. Sources, abundance and fate of gaseous atmospheric pollutants. Final report and supplemental report. Stanford Research Institute, Menlo Park, California.

Ryther, J. H., and W. M. Dunstan. 1971. Nitrogen, phosphorus, and eutrophication in the coastal marine environment. Science 171:1008–1013.

St. Amant, P. P., and L. A. Beck. 1970. Methods of removing nitrates from water. J. Agr. Food Chem. 18:785–788.

Sander, J. 1967. Kann Nitrit in der menschlichen Nahrung Ursache einer Krebsentstehung durch Nitrosaminbildung sein? Arch. Hyg. Bakteriol. 151:22–28.

Sander, J., F. Schweinsberg, and H. P. Menz. 1968. Untersuchungen über die Entstehung cancerogener Nitrosamine in Magen. Hoppe-Seyler's Z. Physiol. Chem. 349:1691–1697.

Sapiro, M. L., S. Hoflund, R. Clark, and J. I. Quin. 1949. Studies on the alimentary tract of the merino sheep in South Africa. XVI. The fate of nitrate in ruminal ingesta as studied *in vitro*. Onderstepoort J. Vet. Sci. 22:357–372.

Sattelmacher, P. G. 1962. Methämoglobinämie durch Nitrat in Trinkwasser. Schriften. Ver. Wasser-, Boden-, Lufthyg. No. 21. Fischer, Stuttgart.

Sawyer, C. N., J. B. Lackey, and A. T. Lenz. 1945. An investigation of the odor nuisance occurring in the Madison lakes, particularly Monona, Waubesa, and Kegonsa, from July 1942–July 1944. Report of Governor's Committee, Madison, Wisconsin. Mimeo.

Scaletti, J. V. 1963. Sublethal effects of nitrate, p. 48. *In* Proceedings of a Conference on Nitrate Accumulation and Toxicity, Cornell University Mimeo 64-6. Department of Agronomy, Cornell University, Ithaca, New York.

Seerly, R. W., R. J. Emerick, L. B. Embry, and O. E. Olson. 1965. Effect of nitrate or nitrite administered continuously in drinking water for swine and sheep. J. Anim. Sci. 24(4):1014–1019.

Sen, N. P., D. C. Smith, and L. Schwinghamer. 1969a. Formation of *N*-nitrosamines from secondary amines and nitrite in human and animal gastric juice. Food Cosmet. Toxicol. 7:301–307.

Sen, N. P., D. C. Smith, L. Schwinghamer, and J. J. Marleau. 1969b. Food additives. Diethylnitrosamine and other *N*-nitrosamines in foods. J. Assoc. Off. Anal. Chem. 52:47.

Shuval, H. I., N. Gruener, S. Cohen, and K. Behroozi. 1970. An epidemiological approach to evaluating the role of nitrates in water in relation to infant methemoglobinemia. *In* H. I. Shuval (ed.) Developments in water quality research. Humphrey Science Pub., Ann Arbor.

Simon, C., H. Manzke, H. Kay, and G. Mrowetz. 1964. Über Vorkommen, Pathogenese und Möglichkeiten zur Prophylaxe der durch Nitrit Verursachten Methamoglobinamie. Z. Kinderheilk. 91:124–138.

Simon, J., J. M. Sund, M. J. Wright, and F. D. Douglas. 1959. Prevention of non-infectious abortion in cattle by weed control and fertilization practices on lowland pastures. J. Am. Vet. Med. Assoc. 135:315–317.

Sinclair, K. B., and D. I. H. Jones. 1967. Nitrite toxicity in sheep. Res. Vet. Sci. 8(1):65–70.

Singley, T. L. 1962. Secondary methemoglobinemia due to the adulteration of fish with sodium nitrite. Ann. Intern. Med. 57:800–803.

Sinha, D. P. 1969. Pathogenesis of abortion in acute nitrite toxicosis in guinea pigs. Diss. Abstr. Int. B. 30(1):268B–269B.

Sinios, A., and W. Wodsak. 1965. Die Spinatvergiftung des Saughlings. Deut. Med. Wochenschr. 90:1856–1863.

Sleight, S. E., and O. A. Atallah. 1968. Reproduction in the guinea pig as affected by chronic administration of potassium nitrate and potassium nitrite. Toxicol. Appl. Pharmacol. 12:179–185.

Smith, G. 1970. The nitrate panic button: What are the facts? Paper presented at Michigan State University Fertilizer Conference, East Lansing. December 4.

Smith, J. E., and E. Beutler. 1966. Methemoglobin formation and reduction in man and various animal species. Am. J. Physiol. 210:347–350.

Smith, S. O., and J. H. Baier. 1969. Report on nitrate pollution of ground water, Nassau County, Long Island. Bureau of Water Resources, Nassau County Department of Health, Mineola, New York.

Sokolovski, V. V., and L. M. Pavlova. 1961. Thiol systems of erythrocytes and the formation of methemoglobin. Biochemistry 25:463–465.

Sokolowski, J. H. 1966. Nitrate poisoning: A chemical and biological evaluation of potassium nitrate utilization by the growing, fattening lamb. Diss. Abstr. 26:6946–6947.

Stallcup, O. T., R. Robberson, O. H. Horton, and R. L. Thurman. 1960. Nitrate poisoning of dairy cattle grazing oat forage. J. Dairy Sci. 43(3):447. (Abstr.)

Statistical Reporting Service, U.S. Department of Agriculture. 1963–1971. Cattle on feed. U.S. Government Printing Office, Washington, D.C.

Stanford, G., C. B. England, and A. W. Taylor. 1970. Fertilizer use and water quality. Agricultural Research Service 41-168, U.S. Department of Agriculture, Washington, D.C. 19 p.

Stauffer, R. S. 1942. Run-off, percolate, and leaching losses from some Illinois soils. J. Am. Soc. Agron. 34:830–835.

Stewart, B. A., F. G. Viets, Jr., G. L. Hutchinson, and W. D. Kemper. 1967. Nitrate and other water pollutants under fields and feedlots. Environ. Sci. Tech. 1:736–739.

Stewart, G. A., and C. P. Merilan. 1958. Effect of KNO_3 intake on lactating dairy cows. Univ. Mo. Res. Bull. 650.

Stewart, K. M., and G. A. Rohlich. 1967. Eutrophication—A Review. Publ. 34. State Water Quality Control Board, State of California, Sacramento.

Stoker, R. E., J. L. Evans, M. C. Stillions, W. V. Chalupa, and C. H. Ramage. 1961. The digestibility of common timothy hays grown with different levels of nitrogen fertilization. J. Dairy Sci. 44:1206.

Stormorken, H. 1953. Methemoglobinemia in domestic animals, Vol. I, part 1, p. 501–506. *In* Proceedings of the Fifteenth International Veterinary Congress. Stockholm, Sweden.

Stout, P. R., and R. G. Burau. 1967. The extent and significance of fertilizer build-up in soils as revealed by vertical distribution of nitrogenous matter between soils and underlying water reservoirs, p. 283–310. *In* N. C. Brady (ed.) Agriculture and the quality of our environment. American Association for the Advancement of Science, Washington, D.C.

Subbotin, F. N. 1961. Nitrates in drinking water and their effect on the formation of methemoglobin. Distributed by the U.S. Department of Commerce

Office of Technical Services, Washington, D.C. Gig. Sanit. 25(2):13–17.
Taylor, A. W., W. M. Edwards, and E. C. Simpson. 1971. Nutrients in streams draining woodland and farmland near Coshocton, Ohio. Water Resource Res. 7:81–89.
Terracini, B., P. N. Magee, and J. M. Barnes. 1967. Hepatic pathology in rats on low dietary levels of dimethylnitrosamine. Brit. J. Cancer 21:559–565.
Thewlis, B. H. 1967. Testing of wheat flour for the presence of nitrite and nitrosamines. Food Cosmet. Toxicol. 5:333–337.
Train, R. E., R. Cahn, and G. J. MacDonald. 1970. Environmental quality. First annual report of the Council on Environmental Quality. U.S. Government Printing Office, Washington, D.C.
U.S. Department of Agriculture. 1963. Agricultural statistics, 1962. U.S. Government Printing Office, Washington, D.C.
U.S. Department of Agriculture. 1969. Agricultural statistics, 1969. U.S. Government Printing Office, Washington, D.C.
U.S. Department of Agriculture. 1970. Agricultural statistics, 1970. U.S. Government Printing Office, Washington, D.C.
U.S. Department of the Interior. 1970a. Agricultural practices and water quality. Proceedings of a conference concerning the role of agriculture in clean water, Iowa State University, November 1969. Federal Water Pollution Control Administration, U.S. Department of the Interior, Washington, D.C.
U.S. Department of the Interior. 1970b. Proceedings, First National Symposium on Food Processing Wastes. Water Pollution Control Research Series 12060-04/70.
U.S. Geological Survey. 1962. Public water supplies of the 100 largest cities in the United States. Water Supply Paper 1812.
U.S. Public Health Service. 1964. Statistical summary of 1962, inventory of municipal waste facilities in the United States. Publ. No. 1165. U.S. Public Health Service, Department of Health, Education, and Welfare. 41 p.
U.S. Public Health Service. 1970. Community water supply study. Analysis of national survey findings. U.S. Public Health Service, Department of Health, Education, and Welfare. 111 p.
University of Wisconsin. 1969. Proceedings of Farm Animal Wastes and By-Products Management Conference. University of Wisconsin, Madison, November 6–7.
Viets, F. G. 1971. Water quality in relation to farm use of fertilizer. Bioscience 21:460–467.
Viets, F. G., Jr., and R. H. Hageman. 1971. Factors affecting the accumulation of nitrate in soil, water, and plants. Agriculture handbook 413. Agricultural Research Service, U.S. Department of Agriculture. U.S. Government Printing Office, Washington, D.C.
Vigil, J., S. Warburton, W. S. Haynes, and L. R. Kaiser. 1965. Nitrates in municipal water supplies cause methemoglobinemia in infant. Public Health Rep. 80:1119–1121.
Volk, G. M. 1956. Efficiency of various nitrogen sources for pasture grasses in large lysimeters of Lakeland fine sand. Soil Sci. Soc. Am. Proc. 20:41–45.
Vollenweider, A. 1968. Scientific fundamentals of the eutrophication of lakes and flowing waters, with particular reference to nitrogen and phosphorus as

factors in eutrophication. Report DAS/CSI/68.27. Organization for Economic Cooperation and Development, Directorate for Scientific Affairs, Paris.

von Kreybig, T. 1965. Die Wirkung einer carcinogenen Methylnitrosoharnstoffdosis auf die Embryonalentwicklung der Ratte. Z. Krebsforsch. 67:46–50.

Vought, R. L. 1966. Epidemiology of goiter, p. 75–85. *In* Proceedings, Second Midwest Conference on the Thyroid. University of Missouri, Columbia.

Wallace, J. D., R. J. Raleigh, and P. H. Weswig. 1964. Performance and carotene conversion in Hereford heifers fed different levels of nitrate. J. Anim. Sci. 23:1042–1045.

Walters, H. 1970. Nitrate in soil, plants, and animals. J. Soil Assoc. 16:3.

Walton, G., 1951. Survey of literature relating to infant methemoglobinemia due to nitrate-contaminated water. Am. J. Public Health 41:986–996.

Weibel, S. R., R. B. Weidner, A. G. Christianson, and R. J. Anderson. 1966. Characterization, treatment and disposal of urban stormwater. Adv. Water Pollut. Res. 1:329–352.

Weiss, S., R. W. Wilkins, and F. W. Hayms. 1937. The nature of circulatory collapse induced by sodium nitrate. J. Clin. Invest. 16:73–84.

Whitt, D. M. 1941. The role of bluegrass in the conservation of the soil and its fertility. Soil Sci. Soc. Am. Proc. 6:309–311.

Winks, W. R., A. K. Sutherland, and R. M. Salisbury. 1950. Nitrite poisoning of pigs. Queensland J. Agr. Sci. 7(1):1–14.

Winter, A. J., and J. F. Hokanson. 1964. Effects of long-term feeding of nitrate, nitrite or hydroxylamine on pregnant dairy heifers. Am. J. Vet. Res. 25(105): 343–361.

Winton, E. F. 1970. The variables in infant methemoglobinemia from nitrate in drinking water. Paper presented at the Twelfth Annual Sanitary Engineering Conference. University of Illinois, Urbana. Feb. 11.

Winton, E. F., R. G. Tardiff, and L. J. McCabe. 1971. Nitrate in drinking water. J. Am. Water Works Assoc. 63:95–98.

Wittwer, S. H., and F. G. Teubner. 1959. The uptake of nutrients through leaf surfaces, p. 235–261. *In* K. Sharrer and H. Linser (ed.) The handbook of plant nutrition and fertilizers. Springer-Verlag, New York.

Wood, R. D., C. H. Chaney, D. G. Waddill, and G. W. Garrison. 1967. Effect of adding nitrate or nitrite to drinking water on the utilization of carotene by growing swine. J. Anim. Sci. 26:510–513.

Wright, M. J., and K. L. Davison. 1964. Nitrate accumulation in crops and nitrate poisoning in animals. Adv. Agron. 16:197–247.

Wuhrmann, K., 1964. Nitrogen removal in sewage treatment processes. Verh. Int. Ver. Limnol. 15:580–596.

Wyngaarden, J. B., J. B. Stanbury, and B. Rapp. 1953. The effects of iodine, perchlorate, thiocyanate and nitrate upon the iodine containing mechanism of the rat thyroid. Endocrinology 52:568–574